A Student Guide to
SPSS

Kendall Hunt
publishing company

Carrie Cuttler
Third Edition

Cover image © Shutterstock, Inc.

Kendall Hunt
publishing company

www.kendallhunt.com
Send all inquiries to:
4050 Westmark Drive
Dubuque, IA 52004-1840

TABLE OF CONTENTS

PREFACE

SPSS stands for Statistical Package for the Social Sciences. As its name implies, the software was developed for use in the social sciences, and it is currently one of the most commonly used statistical software packages in psychology, business, and education. The popularity of SPSS is likely a result of its intuitive drop-down menus that allow for quick and easy computation of some of the most commonly used statistics in the social sciences.

This guide will provide you with easy to follow step-by-step instructions on how to conduct some of the most frequently used analyses in SPSS as well as how to interpret the results. The statistical notation and abbreviations approved for use in the 7th edition of the Publication Manual of the American Psychological Association (informally known as the APA style guide) are introduced, and guidelines on reporting results are consistent with APA style.

Although this guide was originally developed for use in introductory statistics courses, numerous updates have been made to the 2nd and 3rd editions in order to expand the guide's suitability for more advanced courses in statistics (e.g., advanced undergraduate statistics). Independent learners, with a rudimentary knowledge of statistics will also find this a useful introduction to computing, interpreting, and reporting some of the most commonly used statistics in the social sciences.

New to the Third Edition

The 3rd edition of this Student Guide to SPSS has been updated to be compatible with versions 26 and 27 of SPSS. Throughout the guide, elaborations on the meaning and interpretation of the various statistics and demonstrations of more advanced statistical analyses have been added. More specific changes to the various chapters are listed below.

- The notation used for partial correlations (pr) and semipartial correlations (sr) has been updated to be consistent with APA style.
- Chapter Four (which was on simple and multiple regression) has been split into two chapters. Chapter Four now focuses on simple regression, and Chapter Five focuses on multiple regression. Further, information about the standard error of the unstandardized slope (SE_b) has been added to the interpretation and reporting of the results.
- Chapter Five has a new example that describes how to include a nominal predictor variable with more than two categories in a multiple regression analysis as well as how to interpret these results.
- Chapter Six now includes a brief discussion of tolerance statistics.
- Chapter Seven now includes descriptions on how to interpret the common effect size indicators Cohen's d and Hedges' g.
- Chapter Nine has been expanded to include one-way within-groups ANOVA. Further, the description of effect sizes for ANOVA has been simplified to include only partial eta-squared. Eta-squared is no longer considered[1].

---

[1] This is because for one-way between-groups ANOVA eta-squared and partial eta-squared are equivalent.

Note about the Data Used in this Guide

The data used in this guide have mostly been fabricated and are intended only to illustrate the execution of various analyses and interpretation of SPSS output. Although some of the data sets and results mirror actual findings, this is the exception, not the rule, and results of analyses presented herein should not be considered actual scientific evidence.

A LIST OF STATISTICAL SYMBOLS AND ABBREVIATIONS USED IN THIS GUIDE

Symbol/Abbreviation	Meaning
a	Intercept
ANOVA	Analysis of variance
b	Slope (unstandardized)
CI	Confidence interval
d	Cohen's d
df	Degrees of freedom
F	F statistic
g	Hedges' g
HSD	Honestly significant difference (Tukey's)
M	Mean (sample)
M_D	Mean of the difference scores
Mdn	Median
N	Sample size (total sample) or population size
n	Sample size (subsample)
p	p value
pr	Partial correlation
r	Pearson correlation
$r_{YY'}$	Correlation between actual scores and predicted scores
R	Multiple correlation
r^2	Coefficient of determination
R^2	Multiple coefficient of determination
r_{pb}	Point biserial correlation
r_s	Spearman rank order correlation
SD or s	Standard deviation (sample)
SD_D	Standard deviation of the difference scores
s^2	Variance (sample)
SE	Standard error
SE_b	Standard error of the slope
SEE	Standard error of estimate
SEM	Standard error of the mean
sr	Semipartial correlation
sr^2	Squared semipartial correlation
t	t statistic
t_{crit}	Critical value for the t statistic
X	Predictor variable
Y' or $\hat{Y}$	Criterion variable
z	z score

Greek Characters	Meaning
α	Alpha
β	Standardized slope (beta weight)
Δ	Delta or change
η_p^2	Eta-squared
μ	Mean (population)
μ_{lower}	Lower limit of a confidence interval
μ_{upper}	Upper limit of a confidence interval
σ	Standard deviation (population)
σ^2	Variance (population)
ϕ	Phi coefficient

ABOUT THE AUTHOR

© Ryan McLaughlin

Dr. Carrie Cuttler is an Assistant Professor of Psychology at Washington State University (WSU), in Pullman, Washington. She received her Ph.D. in psychology from the University of British Columbia (UBC) and subsequently completed a postdoctoral fellowship in the Department of Psychiatry at UBC.

Dr. Cuttler's primary line of research focuses on elucidating the potentially beneficial and detrimental effects of chronic cannabis use and acute cannabis intoxication. Her current and recent work focuses on examining links between cannabis use and mental health (e.g., depression, anxiety, OCD, ADHD), physical health (e.g., pain), stress, and cognition (e.g., memory, decision-making, executive functioning, creativity, and attention).

Dr. Cuttler has been teaching research methods and statistics for over a decade. This guide was originally developed as part of a project to create a practical laboratory component for UBC's introductory statistics course but has since been adapted to increase its utility for more advanced courses (advanced undergraduate courses). Dr. Cuttler has over 60 publications, including the two previous editions of this book, *A Student Guide to SPSS (1st and 2nd edition)*, and a laboratory guide entitled *Research Methods in Psychology: Student Lab Guide.* She also cowrote an open textbook entitled *Research Methods in Psychology (3rd and 4th Editions)* and edited another open textbook entitled *Essentials of Abnormal Psychology (1st edition).* In her spare time, she likes to travel, hike, bike, run, and watch movies with her husband and son.

An Introduction to SPSS

1

Learning Objectives

In this chapter, you will learn how to create variables, define the properties of those variables, and enter data. You will also learn some handy tools and tricks for manipulating and transforming data.

GETTING STARTED

Opening SPSS

You can open the SPSS program by clicking on the program icon ![icon] or the file name "SPSSStatistics." Note that it is a large program, so it can take a minute for it to open. If you used the default installation, the program will be in your applications (Mac) or program files (PC) in a folder labeled "IBM." When you first open the program, a window like the one shown below will appear. It provides you with several options, including options to run tutorials, open recently opened files, or open a new dataset. For now, select the option in the top left corner to open a **New Dataset** and then click **OK**.

Note that if you want to open an existing dataset that has already been saved on your computer, you can simply click on the data file in your computer menu.[1] Alternatively, you can click on the option to "Open another file" in the window shown below and then locate and open the data file.

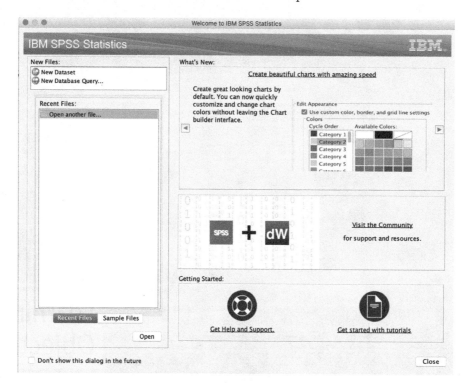

DATA AND VARIABLE VIEW WINDOWS

Data View

Once SPSS is open, the Data View window will appear. As its name implies, this is where you enter the data or where the data will appear if you are opening an existing dataset. Along the top of the screen, you will

---

[1] Trouble shooting tip: If an existing SPSS dataset won't open when you double click on its file name in your computer menu, then try to first open a blank SPSS spreadsheet using the "New Dataset" option. Once a blank SPSS spreadsheet is open, go to File →Open →Data (or click the orange folder in the icon toolbar), locate the file on your computer, highlight the file, and then click Open.

see a toolbar with different options (e.g., File, Edit, View, Data, Transform, Analyze) and another toolbar with a series of different icons.

For now, let's focus on the matrix of cells occupying most of the screen. Along the top of this matrix, you will see a row of cells each labeled "var" to represent variable. Along the left side of this matrix, you will see a column of numbered cells. The columns in Data View are used to designate the different variables. The rows in Data View are used to designate the different participants (or cases) assessed. Thus, any given cell represents a specific participant's score on a specific variable.

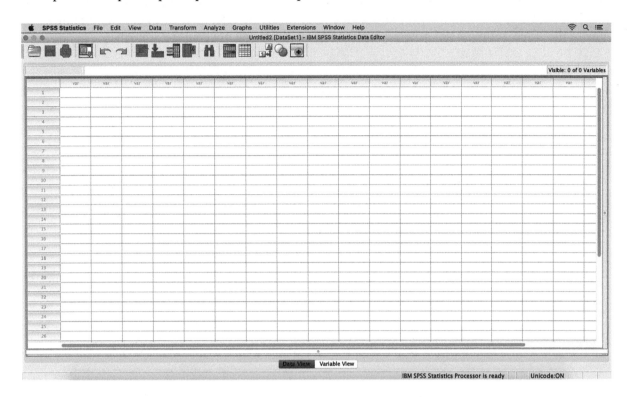

You can highlight cells in the matrix by clicking on them. You can use the arrow keys to move around to different cells. Pressing enter will move you down one cell. Take a second now to practice moving around the cells using your track pad, mouse, and/or arrow keys.

At the bottom of the screen, you will see a tab labeled "Data View" and a tab labeled "Variable View." As shown in the image displayed above, the tab labeled Data View is currently highlighted because you are looking at the Data View window. Click the **Variable View** tab to open the Variable View window.

Variable View

The Variable View window is used to enter the names of the variables and to define the properties of those variables. In the case of an existing dataset, it can be used to learn about the properties of the different variables. As described fully below, the variable names need to be entered in the first column, and their properties are entered in the remaining columns. Once the variable names are entered in the first column of the Variable View window, they will appear in the top row of the Data View window in place of "var." Note that the display in Variable View is transposed from the one in Data View; the Variable View window shows the variable names in the first column, while the Data View window shows the variable names in the top row.

The column labeled "Name" is where you need to enter the names of your variables. The column labeled "Type" is where you can indicate the type of data you will enter (e.g., numeric, currency, string). The column labeled "Width" is where you can set the maximum number of characters that the values of the variables can contain. The column labeled "Decimals" is where you can set the number of decimal places you want displayed for the values of the variables. The column labeled "Label" is where you can enter text descriptions of the variables. The column labeled "Values" is where you can define any codes you are using to identify the values of the variables (e.g., 1 = Men). The column labeled "Missing" permits you to enter any special code (e.g., −99) that you used to identify missing values. The column labeled "Align" can be used to indicate how you would like the data in the Data View window to be aligned (right, left, or center justified). The column labeled "Measure" is where you can specify the scales of measure used to measure the variables (e.g., nominal). Note SPSS uses the term "Scale" to refer to both interval and ratio scales. Finally, the "Role" setting is not generally used, and as such we will not consider it further.

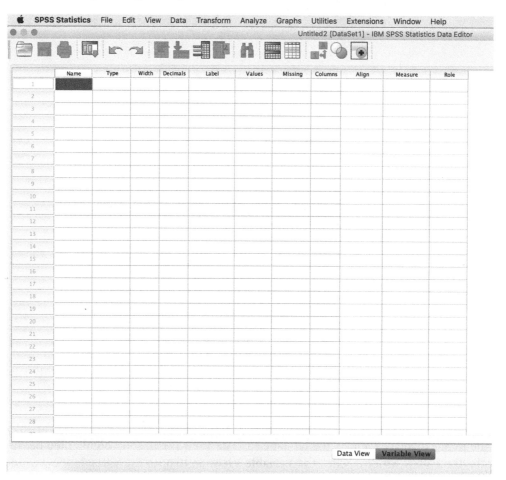

Creating Variables

Let's get started by trying to create some variables. We will start by creating a participant ID variable. While you are in the Variable View window, type **ID** in the first cell in the **Name** column (the cell that is highlighted blue in the image displayed above). Variable names must be one word; spaces and special characters like / and - are not allowed (but _ is permitted). An error message stating that the variable name contains an illegal character will appear if you attempt to use one of these forbidden characters. The ID codes we will use will be numeric, and they will be under 8 digits long so you can leave the Type as the default Numeric and the Width at the default of 8 characters. Since we do not want our ID codes to contain a decimal remainder, we will need to change the default of 2 to 0. To do this you will need to move over to the **Decimals** column and **click on the down arrow**, which appears when you click on this cell, until a value of 0 appears. Our variable name is quite informative, but this is not always the case, so we should practice entering a label for the variable. Click on the first cell in the **Label** column and type **Participant ID Codes**. Finally, set the scale of measure to nominal by moving over to the **Measure** column and clicking on the option for **Nominal.** Now our first variable has been created, and its properties have been defined!

Next let's create a weight variable to represent the weight of each participant in pounds. Type the word **Weight** into the second cell in the **Name** column (immediately under ID). The weights we will enter will be in numerical format, so leave the Type as Numeric. Once again, by default the value in the Width column is set to 8, and the value in the Decimals column is set to 2. Since the weights we will enter will be under 8 digits long, you do not need to adjust these settings. To provide a meaningful description of the variable, you should type **Weight in Pounds** in the second cell in the **Label** column. Finally, since weight is measured on a ratio scale, you should set the scale of measure to scale (the term SPSS uses for interval and ratio scales), by clicking on the relevant cell in the **Measure** column and then clicking on the word **Scale.**

Let's also create a gender variable. Type the word **Gender** into the third cell in the **Name** column (immediately under Weight). We will use numerical codes to represent the different genders, so you should leave Type as Numeric. We will change the width of the cells to 1, but before we can do so we need to change the decimals to 0. Start by clicking on the relevant cell in the **Decimals** column, and then click on the **down arrow button** until a value of **0** is displayed. Now we can change the width of the cells to 1 by highlighting the relevant cell in the **Width** column and clicking the **down arrow button** until it displays a value of **1**. To label the variable, type **Gender** in the third cell in the **Label** column.

SPSS will not permit analyses on string variables (variables containing text), so if we want to perform any analyses with the gender variable, we will need to use numeric codes to label the genders. We will use the number 1 to designate men and the number 2 to designate women. To assign these numerical codes, move over to the adjacent cell in the **Values** column and click the little **grey box with three dots** that appears when the cell is highlighted. A "Value Labels" dialogue window, like the one shown below, will now appear. Enter a **1** in the **Value** box and **Men** in the **Label** box. Click **Add**. You just told SPSS that the gender code 1 means men. Now enter **2** in the **Value** box and **Women** in the **Label** box. Click **Add**. Now SPSS knows that the gender code 2 means women. Click **OK** to close the dialogue window. Finally, change the scale of measure to nominal by clicking on the relevant cell in the **Measures** column and then clicking on the word **Nominal**.

Let's create one last variable. Type the word **Ethnicity** into the fourth cell in the **Name** column (immediately under Gender). By default, SPSS permits only numbers to be entered in the Data View spreadsheet. To change this default setting you need to click on the relevant cell in the **Type** column and then click the little **grey box with three dots** that appears when the cell is highlighted. A "Variable Type" dialogue window, like the following, will now appear. Click on the **String** option, and then click **OK** to close the dialogue window. Now we will be able to enter ethnicities using text rather than numerical codes. Changing type to string automatically changes the decimals to 0, the alignment to left, and the measure to nominal (since any data that contains letters is likely nominal/categorical in nature). Change the width of the cells to 9 by clicking on the relevant cell under the **Width** column and then clicking the **up-arrow button** until it displays a value of **9**. To label the variable, type **Ethnicity** in the third cell under the **Label** column. Since we will be using text rather than codes to enter participants' ethnicities, there is no need to define the values of the variables.

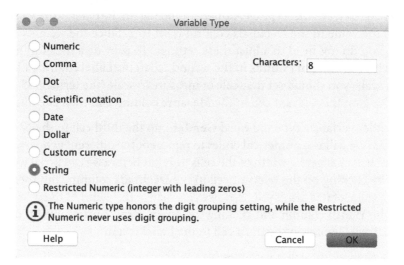

Your Variable View window should now look like the one shown here.

	Name	Type	Width	Decimals	Label	Values	Missing	Columns	Align	Measure	Role
1	ID	Numeric	8	0	Participant ID C...	None	None	8	Right	Nominal	Input
2	Weight	Numeric	8	2	Weight in Pounds	None	None	8	Right	Scale	Input
3	Gender	Numeric	1	0	Gender	{1, Men}...	None	8	Right	Nominal	Input
4	Ethnicity	String	9	0	Ethnicity	None	None	8	Left	Nominal	Input

Now return to the Data View window by clicking on the **Data View** tab at the bottom of the screen. Voilà! As shown below, the variable names we created now appear in the top row of the Data View window.

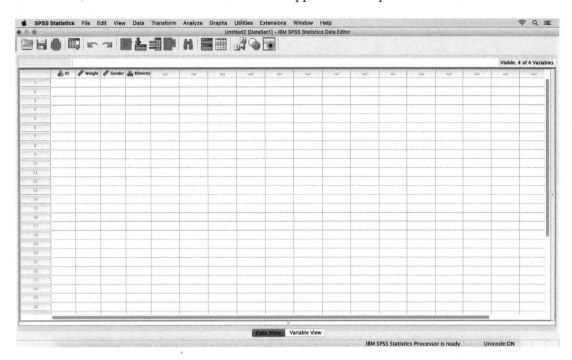

Entering Data

Recall that in the Data View window each row represents a single participant's scores on each of the variables. As such, we will need to enter the participant ID codes into the 1st column, the participants' weights into the 2nd column, the participants' gender codes into the 3rd column, and the participants' ethnicities into the 4th column. Try entering the following data into the spreadsheet by clicking on the relevant cell and typing in the information. Do not enter the variable names shown below in the top row, as they should already appear in the top row of your Data View spreadsheet.

ID	Weight	Gender	Ethnicity
1	105	2	Asian
2	165	1	Caucasian
3	135	2	Caucasian
4	185	1	Caucasian
5	120	2	Asian

Your screen should now look like the one shown here. It should now be clear that each row represents a different participant and each column represents a different variable. Thus, each cell represents one participant's score on one of the variables. The first participant's ID code is 1. That participant weighs 105 lb, is a woman (which we coded 2), and is Asian. The second participant's ID code is 2. That participant weighs 165 lb, is a man (which we coded 1) and is Caucasian. The third participant's ID code is 3. She is a Caucasian woman weighing 135 lb. The fourth participant's ID code is 4. He is a Caucasian man weighing 185 lb. The last participant's ID code is 5. She is an Asian woman who weighs 120 lb.

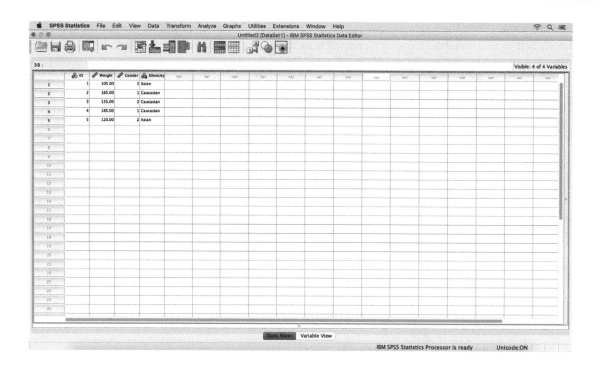

SAVING THE SPSS DATASET (FILE→SAVE AS)

Now that we've done all of this work, we want to make sure we don't lose it (we will use this dataset in the remainder of this chapter as well as in Chapter Two). You can save the file either by pressing the **disk icon** in the icon toolbar or by clicking **File→Save As.** When prompted, type in the file name **Practice Data**, find a good location on your computer to save it and press **Save**. An output window will appear, displaying your command to save the dataset. You can simply close that window for now. A prompt to save the output window will appear when you try to close it. Since we only want to save the dataset (and not the code generated to save the data), you can click **No**.

> Congratulations you now know how to open SPSS, create variable names, define the properties of the variables, enter data and save your data. You're off to a great start!

HANDY TOOLS AND TRICKS

Viewing Value Labels

Typically, in order to decode variable codes (e.g., gender codes), you need to look them up in the Values column in the Variable View window. However, a shortcut is available, allowing you to quickly view variable codes in the Data View window. This is useful if you forget what your codes represent or if you are using a dataset that someone else prepared and you do not know what the various codes represent. To decode variables that are coded, press the **Value Labels icon** in the icon toolbar. It is the icon in the image shown here. Once you click this icon, you will see that all of the gender codes (the 1s and 2s) change to the labels of the codes we previously entered in the Value Labels dialogue window (Men and Women). To recode the variable, simply press the icon again.

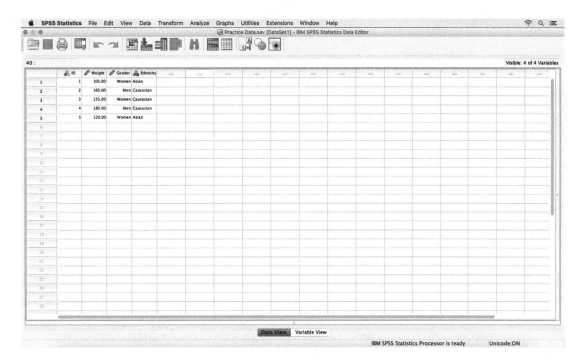

Sorting Data (Data→Sort Cases)

It is often useful to sort your data according to the values of a variable (i.e., to reorder scores from lowest to highest or highest to lowest). For instance, if you wanted to find the most extreme scores on a variable, you could sort them and then look at the scores at the very top and very bottom of the column. Let's practice using this function by sorting the data according to weight. Go to the upper tool bar and click **Data→Sort Cases**.

© Photoraidz/Shutterstockcom

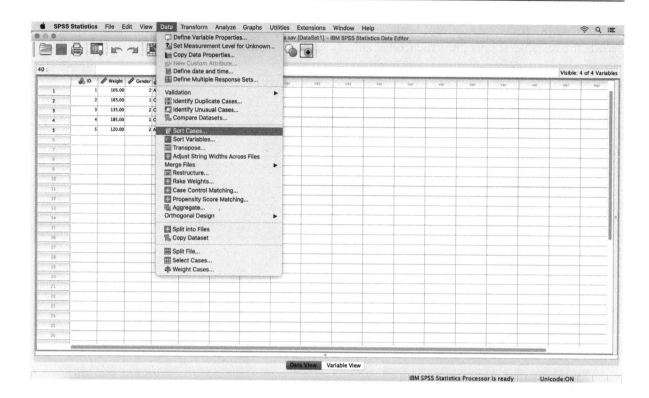

A "Sort Cases" dialogue window, like the one displayed below, will now appear. Highlight the **Weight** variable, shown on the left side of the dialogue window, by clicking on it. Once it is highlighted, click the **blue arrow** to move it over to the box on the right labeled "Sort by:". Notice you can choose to sort from lowest to highest (ascending) or from highest to lowest (descending). Leave it at the default, which is to sort in ascending order, and press **OK** to close the window.

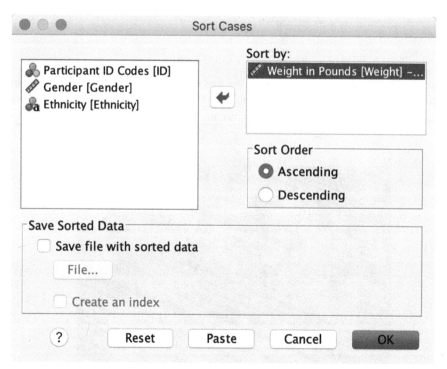

An output window will appear, displaying the command to sort the data. You can simply close that window. Once again, there is no need to save the output window since it contains only the code for the command, so select **No** when prompted to save the output.

Notice that the data are now organized such that the lightest participant's data (ID 1) are in the first row and the heaviest participant's data (ID 4) are in the last row. You should practice using this feature by re-sorting the data, so they are once again organized according to ID code.

Selecting Cases (Data→Select Cases)

Another handy tool is selecting cases. This tool allows you to perform calculations and analyses using only a specific subset of the participants. For example, if you wanted to know the average weight of only the men, you would first need to select only the men's data before computing the mean weight (computation of the mean is discussed in Chapter Two). Let's practice by selecting only the men's data. Go to the upper toolbar and click **Data→Select Cases**. Alternatively, you can simply click on the **Select Cases icon** in the icon toolbar. It is the icon shown below.

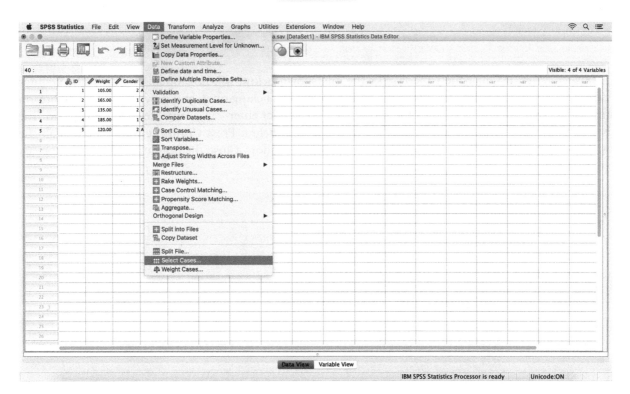

A "Select Cases" dialogue window, like the one displayed here, will now appear. The default is to have all cases selected (you can think of cases as participants). To select only a subset of the cases, select **If condition is satisfied** and press the **If... tab**.

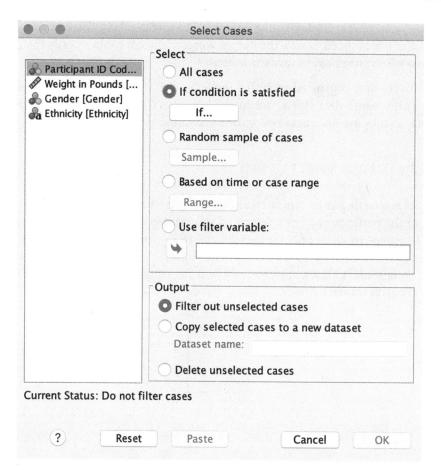

A "Select Cases: If" dialogue window, like the one shown below, will now open. Highlight the **Gender** variable located on the left side of the dialogue window by clicking on it, and move it over to the box on the upper right by clicking on the **blue arrow**. Next to Gender enter **= 1,** using the onscreen keypad. The box should now read Gender = 1, like the one displayed below. Press **Continue** to close this dialogue window**.** Press **OK** to close the "Select Cases" dialogue window. Close the output window that appears without saving.

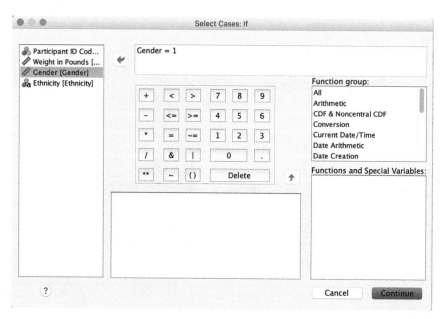

As shown in the image displayed below, lines will now appear through the cells next to the three women. The lines indicate that these participants will be dismissed or excluded from future analyses. You should also notice that a new column labeled "filter_$" has been created (highlighted blue in the image shown below). The 0s in that column indicate the participants who will be excluded from subsequent analyses because they don't meet the criteria you specified (women), and the 1s indicate the participants who will be included in subsequent analyses because they do meet the criteria you specified (men). Once again, you can also reveal these codes by clicking the Value Labels icon in the icon toolbar. If you use this option, then the words "Selected" and "Not Selected" will appear rather than 1s and 0s.

Note that if you wanted to perform subsequent analyses on the selected cases alone, you would still use the variable of interest when performing the analysis. For instance, if you wanted to calculate the mean weight of only the men, you would set the filter to select only men, and then you would calculate the mean, using the weight variable. Never attempt to conduct an analysis using the "filter_$" variable as a variable. The column is created only to indicate which participants have been filtered out; it is not a real variable and cannot be analyzed in any meaningful way.

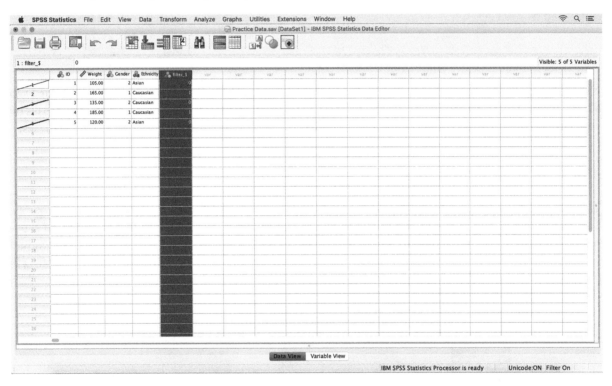

To reselect all cases, you can simply delete the column labeled "filter_$" by **right clicking** on the **column header** (the filter_$ label) and clicking **Cut**. Alternatively, you can click on the **Select Cases icon** in the icon toolbar or go to **Data→Select Cases** and then, using the "Select Cases" dialogue window, click **All Cases** and then press **OK**. If you use this alternative option, you will notice that the lines through the cells disappear but the "filter_$" column remains (although it is now inactive).

Next, let's try selecting participants with weights less than or equal to 165. Click on the **Select Cases icon** in the icon toolbar or go to **Data→Select Cases.** Select **If condition is satisfied** and press the **If…** button. Delete any information in the upper white box (Gender = 1 should still be in there), highlight the **Weight** variable on the left side of the dialogue window by clicking on it, and move it into the box on the top right side by clicking on the **blue arrow**. Next to Weight enter **<= 165,** using the onscreen keypad (note: <= designates less than or equal to, while => designates greater than or equal to). Your screen should now look like the one displayed below. Click **Continue** and then click **OK** to close the dialogue windows. Close the output window without saving the output.

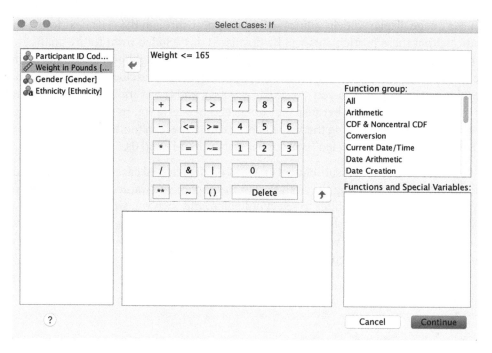

Now the participant with a weight over 165 will not be entered into any future analyses, and your Data View window will show a line through the cell next to his participant ID. Try reselecting all of the cases on your own so that this participant will not be excluded from our subsequent computations and analyses.

Splitting Data into Groups (Data→Split File)

The split file option is useful when you want to perform separate analyses on subsets of the participants. This tool allows you to split your file into subgroups of participants so that they can be considered separately. For instance, if we wanted to know the average weight of men and the average weight of women separately, we could use this option to split our file by gender before computing the mean weight. Let's practice using this tool by splitting the dataset by gender. Go to **Data→Split File**. Or using the icon toolbar, click on the icon shown below.

© sibgat/Shutterstock.com

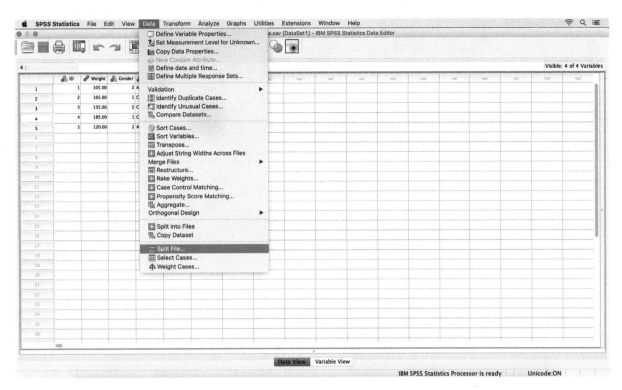

A "Split File" dialogue window will appear. Click **Organize output by groups,** and then move the variable **Gender** to the **Groups Based on: box** by highlighting the variable name **Gender** and then clicking on the **blue arrow**. Click **OK**. Close the output window without saving the output.

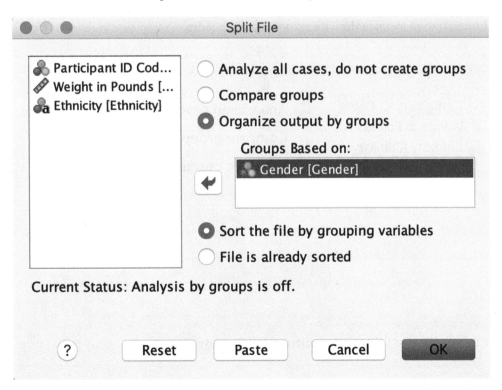

As shown in the following display, the data are now sorted by gender. In addition to the data being reorganized, any subsequent analyses will now be performed separately for men and women, and the results of these analyses will be presented for each of these groups separately.

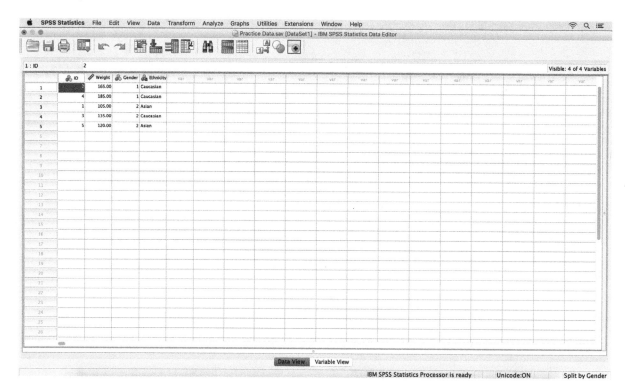

Since we do not want our dataset to remain split, we should deactivate this feature by going to **Data→Split File** and selecting **Analyze all cases, do not create groups**. Click **OK** and then close the output window without saving the output. Notice that the data remain organized according to gender, but any subsequent analyses will be performed on the entire sample. This is a good opportunity for you to practice using the sort cases function to reorganize the data according to ID codes.

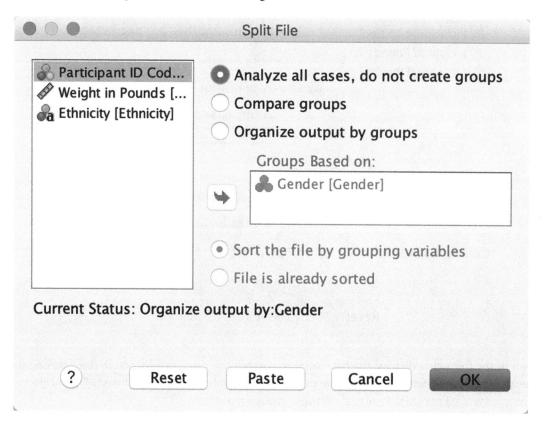

Recoding Variables (Transform→Recode into Different Variables)

Sometimes, we need to recode a variable. For instance, since we entered the participants' ethnicities as a string (text) variable, SPSS will not allow us to perform any analyses with the variable. In order to perform analyses on this variable, we will need to recode it using numerical codes. To recode an existing variable into a different variable, go to the top toolbar and click **Transform→Recode into Different Variables.**[2]

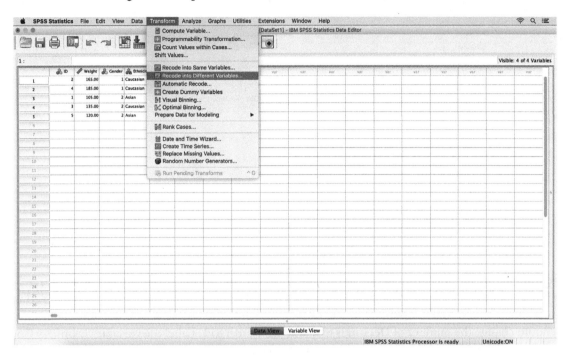

A dialogue window labeled "Recode into Different Variables," like the one shown below, will now appear. Enter the variable you want to recode—in this case **Ethnicity**—into the box labeled **String Variable – > Output Variable** by highlighting the variable name in the box on the left and clicking on the **blue arrow**. Enter the name of the new variable you want to create—in this case **EthnicityCode**—in the box labeled **Name**. Click on the **Change tab** highlighted blue in the image shown below.

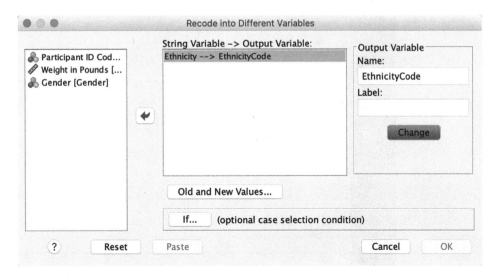

---

[2] We could also use the option to "Recode into Same Variables," but this would overwrite our original variable.

Next click the tab labeled **Old and New Values**. A new dialogue window, like the one shown here, will now appear. We will recode "Caucasian" as "1" and "Asian" as "2." To do this, you will need to enter **Caucasian** into the **Old Value: Value box** and enter a **1** into the **New Value: Value box.** Click **Add.** Next enter **Asian** into the **Old Value: Value box** and enter a **2** into the **New Value: Value box.** Click **Add.** Finally, click **Continue** and then **OK** to close the dialogue windows. Finally, close the output window that will appear without saving it.

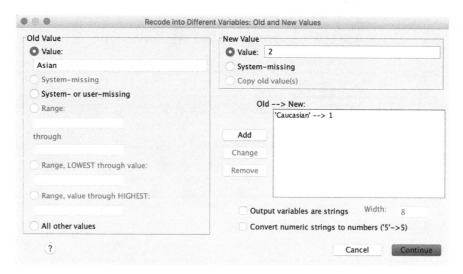

As highlighted in the following image, a new variable labeled "EthnicityCode" will now appear in the Data View window. Caucasian participants are labeled with a 1, and Asian participants are labeled with a 2. You should now go to the Variable View window and define the properties of this new coded variable. You should change the number of **Decimals** to **0** and the **Width** to **1**. Label the variable "Ethnicity Coded," define the values of this new coded variable using the Values column, and define the scale of measure as **Nominal**, using the **Measures** column. You should refer to the section entitled "Creating Variables (in Variable View)" in this chapter if you forget how to define the properties of variables.

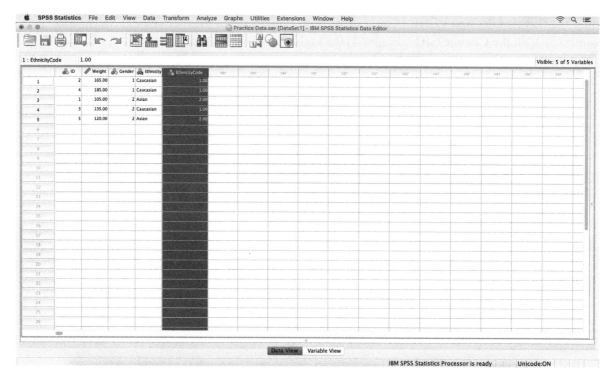

Computing New Variables (Transform→Compute Variable)

Sometimes, we need to compute a new variable by transforming an existing variable. For instance, let's say we wanted to compute all of the participants' weights in kilograms. To compute a new variable using an existing variable, go to the top toolbar and click **Transform→Compute Variable**.

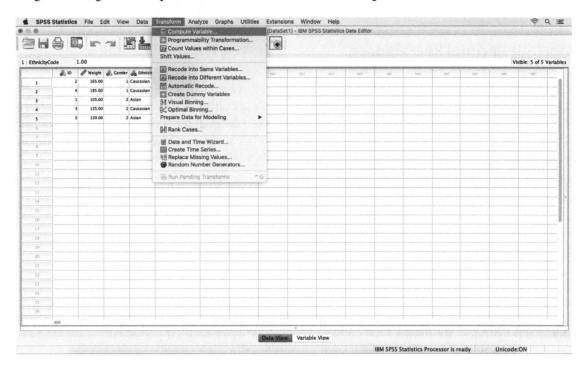

A "Compute Variable" dialogue window will now appear. Type the name of the new variable you want to create—**Kg**—in the **Target Variable box**. Highlight the **Weight** variable on the left side of the dialogue window, and enter it into the **Numeric Expression** box on the upper right side using the **blue arrow button**. One lb. equals 0.45 kg, so to convert all of the weights from pounds to kilograms, you will need to multiply each weight by 0.45. To do this, enter *****.45** next to the variable name Weight (the * indicates multiplication, a / would indicate division). Your screen should now look like the one shown below. Press **OK** to close the dialogue window. Close the output window without saving.

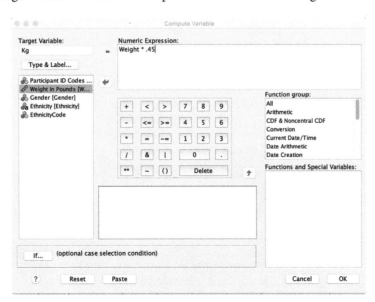

A new variable labeled "Kg" will now appear in your Data View window, with all of the participants' weights in kilograms. You can do just about any mathematical transformation to your data using this handy tool!

REPORTING DECIMAL REMAINDERS AND ROUNDING

Reporting Decimal Remainders

The 7th edition of the *Publication Manual of the American Psychological Association* (the APA style guide) advises us to report our results to two decimal places. There are only a few exceptions to this general rule. First, discrete variables, which are variables that cannot have a decimal remainder (e.g., number of participants), should be presented with no decimal remainder. Also, *p* values between .001 and .009 should be reported to three decimal places, while those lower than .001 should be reported as $p < .001$.

Statistics that cannot take on values greater than 1 (e.g., correlation coefficients, *p* values) should be reported without a leading 0 before the decimal place (e.g., $p = .03$). Statistics and other values that can be greater than 1 (e.g., standard deviations, variance) should be reported with a leading 0 before the decimal place when the values are lower than 1 (e.g., $s^2 = 0.36$).

Rounding

The results of analyses provided in the SPSS output will be rounded to various decimal places; sometimes no decimal remainders are shown, while other times five or more decimal places are provided. As such, some rounding is typically required. If you want to round to two decimal places and the value in the third decimal place is lower than 5, you should round down, but if the value in the third decimal place is a 5 or a value higher than 5, you should round up. When a table in the output window shows a value rounded to three decimal places and the number in the third decimal place is a 5, you will need to determine if the value has been rounded up or down. To do so, simply click on the value in the output window several times until the entire string of digits after the decimal place is displayed. You will then be able to make a decision about whether you need to round the number in the second decimal place up or down.

Descriptive Statistics

Learning Objectives

In this chapter, you will learn how to calculate and report indicators of central tendency (mean, median, mode), variability (range, standard deviation, variance), and the shape of a distribution (skewness, kurtosis). You will also learn how to transform a set of raw scores to z scores.

CENTRAL TENDENCY AND VARIABILITY

Most of the analyses you will need to conduct can be found in the upper toolbar under the drop-down menu labeled "Analyze." Options to compute descriptive statistics and to analyze data using correlation, regression, *t*-tests, and other important statistics can be found there. We'll start with computing indicators of central tendency and variability.

Computing Indicators of Central Tendency and Variability (Analyze→Descriptive Statistics→Frequencies)

Let's compute descriptive statistics on the weight variable we created in Chapter One. Open the practice data set you created in Chapter One by clicking on the file name in your computer menu. Once the file is open, use the upper toolbar to go to **Analyze→Descriptive Statistics→Frequencies**.

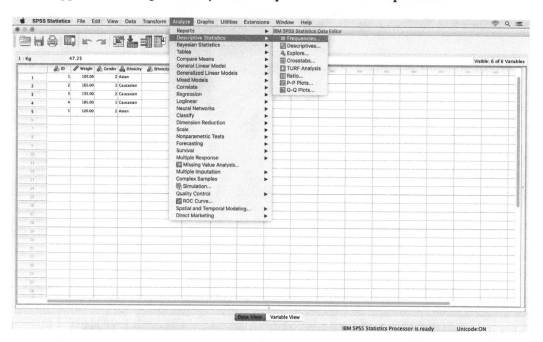

A "Frequencies" dialogue window, like the one shown below, will now open. Highlight the variable **Weight in Pounds** shown on the left side of the window by clicking on it, and press the **blue arrow** to move it into the box labeled **Variable(s)**.

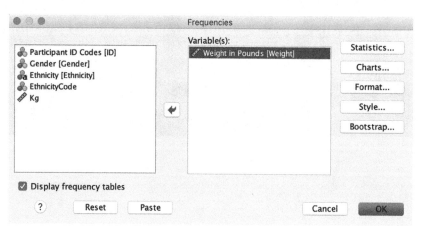

Next, press the **Statistics… tab** in the "Frequencies" dialogue window. A "Frequencies: Statistics" dialogue window, like the one shown below, will now open. Check the boxes next to **Std. deviation, Variance, Range, Minimum, Maximum, Mean, Median, and Mode**.

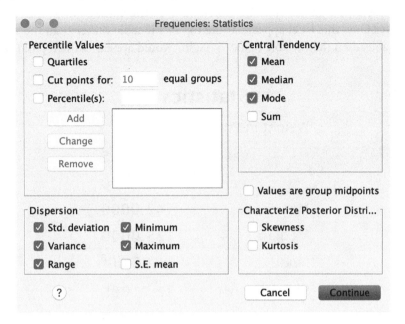

Press **Continue** and then **OK** to close the dialogue windows.

Interpreting the Results

An output window, like the one shown below, will now appear. Whenever you analyze data in SPSS, the results will appear in an output window like this one.

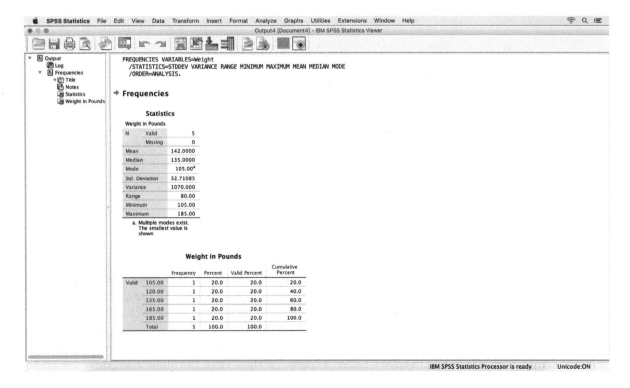

Descriptive Statistics

Let's take a closer look at the "Statistics" table shown below. As you can see, the table displays descriptive statistics for the variable "Weight in Pounds." Since we gave the variable this label when we defined its properties in Chapter One, this label appears at the top of the table rather than the variable name. If we didn't label the variable, then the variable name "Weight" would appear at the top of the table.

Statistics

Weight in Pounds

N	Valid	5
	Missing	0
Mean		142.0000
Median		135.0000
Mode		105.00[a]
Std. Deviation		32.71085
Variance		1070.000
Range		80.00
Minimum		105.00
Maximum		185.00

a. Multiple modes exist. The smallest value is shown

Sample Size (N)

The row labeled "N" shows that the size of the sample is 5. The value of 5 next to the label "Valid" indicates that we have data on the weights of 5 participants. The value of 0 next to the label "Missing" indicates that we are not missing any data on participants' weights. If you deleted the weight of one of the participants and reran the analysis, a value of 4 would appear next to "Valid," and a value of 1 would appear next to "Missing" to indicate that you are missing data on one participant's weight.

Variability

The remainder of the table displays the indicators of central tendency and variability that we requested. We will start with the indicators of variability, since they are the most straightforward to interpret. The row labeled "Std. Deviation" shows the standard deviation is 32.71. In other words, the average distance of scores from the mean is 32.71 units. The row labeled "Variance" shows the variance is 1070.00 (the variance is just the standard deviation squared). The row labeled "Range" shows the range of weights is 80.00; that is, the highest weight is 80 lb. higher than the lowest weight. The row labeled "Minimum" shows that the lowest weight is 105.00 lb, and the row labeled "Maximum" shows that the highest weight is 185.00 lb.

Central Tendency

Next, we will proceed to the indicators of central tendency. The row labeled "Mean" shows the mean (i.e., average) weight is 142.00. The row labeled "Median" shows that the median (i.e., middlemost) weight is 135.00. Importantly, you should note that the "Statistics" table does not provide reliable information on modes. If the data contains one mode, the table will display its proper value. However, if there are multiple modes, the table will show only the lowest mode, and it will put a note under the table (like the one shown in the "Statistics" table displayed previously) to let you know that multiple modes exist and only the smallest value is shown. The mode is just the most frequently occurring score, so to find the other modes you can simply refer to the frequency table that also appears in the output window. This table is presented below.

Weight in Pounds

		Frequency	Percent	Valid Percent	Cumulative Percent
Valid	105.00	1	20.0	20.0	20.0
	120.00	1	20.0	20.0	40.0
	135.00	1	20.0	20.0	60.0
	165.00	1	20.0	20.0	80.0
	185.00	1	20.0	20.0	100.0
	Total	5	100.0	100.0	

The table clearly shows each value of the variable (each weight) in the first (unlabeled) column. The column labeled "Frequency" shows the frequency with which each of the values occurred (each of the weights occurred only once, so 1 is listed in each cell). The column labeled "Percent" shows the relative frequency (i.e., the percentage of participants with each score), and the column labeled "Cumulative Percent" lists the cumulative percentages (the percentage of participants with each score or a score lower).

When two modes exist, we say the distribution is bimodal, and we report both modes, but when all of the scores in a distribution have the same frequency of occurrence, we report that there is no mode. As shown above, the scores in the weight distribution all have the same frequency (a frequency of 1), so while the "Statistics" table in the output window indicates that there are multiple modes, you would need to report that there is no mode.

Reporting the Results

According to the American Psychological Association (APA) style guidelines, most statistical notation should be italicized (the exceptions are Greek characters like μ and subscripts). An uppercase N is used to denote the size of the population, whereas a lowercase n is typically used to denote the number of participants in a sample. However, an uppercase N can also be used to denote the total number of participants in a sample, while a lowercase n can be used to denote the number of participants in a subsample (e.g., number of men in the sample).

According to the APA style guide, the mean of a population is symbolized using μ, while the mean of a sample is symbolized using M. The statistical notation endorsed by the APA for the median is Mdn. The notation s^2 should be used for the variance, but either s or SD can be used to denote the standard deviation. Once again, you should defer to your instructor for guidelines on which of these two notations to use for the standard deviation. The APA style guide does not provide abbreviations for the mode or the range. Moreover, while technically the range is the highest score minus the lowest score, when we report the range using APA style, we simply report both the lowest and highest scores separated by a dash.

So, using APA style we would report the size of the sample as $N = 5$. If we wanted to report the number of men (a subsample), we could report, $n = 2$. For indicators of central tendency, we would report: $M = 142.00$, $Mdn = 135.00$, and no mode. For the indicators of variability, we would report: $SD = 32.71$, $s^2 = 1070.00$, and range $= 105.00 - 185.00$. Note that since the words mode and range are not considered notation, they do not need to be italicized.

SKEWNESS AND KURTOSIS

Skewness

Skewness and kurtosis statistics can help give you a sense of the shape of your distribution and allow you to determine whether your scores are indeed normally distributed. The skewness statistic provides an indicator of the degree of asymmetry in your distribution. A skewness statistic of 0 would indicate a perfectly symmetrical distribution. Values above 0 (positive values) indicate that the distribution is positively skewed. As shown in the figure below, a positively skewed distribution contains more low scores (i.e., a peak on the left side of the distribution) and a longer right tail. Values below 0 (negative values) indicate that the distribution is negatively skewed. As shown in the figure below, a negatively skewed distribution contains more high scores (i.e., a peak on the right side of the distribution) and a longer left tail.

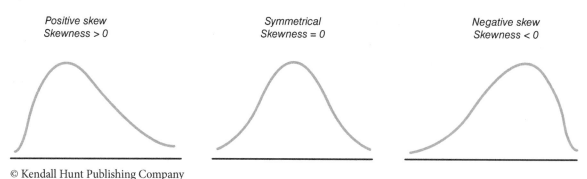

| Positive skew | Symmetrical | Negative skew |
| Skewness > 0 | Skewness = 0 | Skewness < 0 |

© Kendall Hunt Publishing Company

Kurtosis

The kurtosis statistic provides an indicator of how tall and narrow the central peak is and how fat the tails of the distribution are, relative to the normal distribution. A perfectly normal distribution has a kurtosis value of 3. However, many statistics programs, including SPSS, provide "excess kurtosis" values, which is the standard kurtosis value minus 3. Thus, SPSS would report an excess kurtosis value of 0 for a perfectly normal distribution. Distributions with excess kurtosis values of 0 are referred to as mesokurtic. Distributions with excess kurtosis values above 0 (positive values) are referred to as leptokurtic. As shown in the image displayed here, leptokurtic distributions have a taller narrower central peak and tails that are fatter (raised). Distributions with excess kurtosis values below 0 (negative values) are referred to as platykurtic. As shown below, platykurtic distributions have a flatter central peak and tails that are shorter and thinner (remember: platy = flatty).

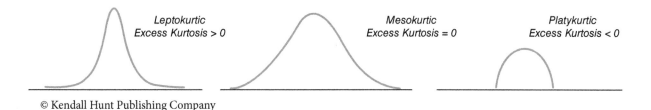

© Kendall Hunt Publishing Company

Computing Skewness and Kurtosis Statistics (Analyze→Descriptive Statistics→Frequencies)

Skewness and kurtosis statistics are also considered descriptive statistics, so they can also be computed by using the upper toolbar to go to **Analyze→Descriptive Statistics→Frequencies**. The "Frequencies" dialogue window will reappear. The weight variable should already be in the Variable(s) box from the previous analyses (if it is not, then use the blue arrow to move it over). Next, click on the **Statistics tab**. The "Frequencies: Statistics" dialogue window shown below will now appear. Check the boxes next to **Skewness** and **Kurtosis** (you may wish to uncheck the remaining boxes to keep your output window less cluttered). Press **Continue** and then **OK** to close the dialogue windows.

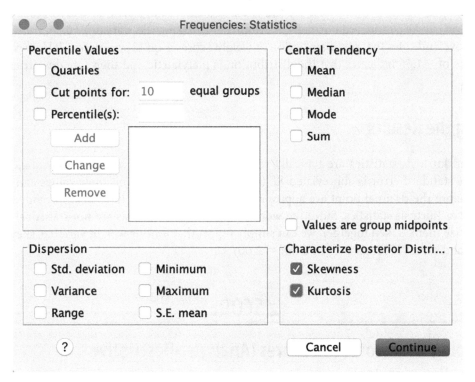

© Kendall Hunt Publishing Company

Interpreting the Results

The "Statistics" table shown here will now appear in your output window. The row labeled "Skewness" provides the value of the skewness statistic. As you can see, it is 0.36. Similarly, the row labeled "Kurtosis" displays an excess kurtosis statistic of −1.66. The table also presents the values of the standard error for the skewness and kurtosis statistics in the rows labeled "Std. Error of Skewness" and "Std. Error of Kurtosis," respectively.

Statistics

Weight in Pounds

N	Valid	5
	Missing	0
Skewness		.357
Std. Error of Skewness		.913
Kurtosis		−1.657
Std. Error of Kurtosis		2.000

As described previously, perfectly normal distributions have skewness and excess kurtosis values of exactly 0. Since few things in life are ever perfect, these values will typically deviate from 0. The further the skewness and excess kurtosis values deviate from 0, the more the distribution deviates from normality. As a crude but general rule of thumb, if the skewness and excess kurtosis values are between −1 and +1, the distribution can be considered normally distributed.[1] The skewness statistic of 0.36, shown in the table displayed above, indicates that our distribution of weight scores is quite symmetrical. However, the excess kurtosis value of −1.66 indicates that the distribution is platykurtic and therefore deviates slightly from normality.

Reporting the Results

Skewness and kurtosis statistics are typically reported along with the value of their standard error in brackets. The standard error is abbreviated *SE* (note the italics used), and their values can exceed 1 (so leading 0s before the decimal point are appropriate). No abbreviations have been provided by the APA for skewness or kurtosis statistics. Since the words skewness and kurtosis are not statistical notation, we will not italicize them. Taken together, we can simply report that for our weight variable, skewness = 0.36 (*SE* = 0.91) and excess kurtosis = −1.66 (*SE* = 2.00).

z SCORES

Transforming Raw Scores to z Scores (Analyze→Descriptive Statistics→Descriptives)

To transform a set of raw scores to *z* scores (i.e., standardized scores), use the upper toolbar to go to **Analyze→Descriptive Statistics→Descriptives**. Note that this option can also be used to compute descriptive statistics; however, it does not include options to compute the median or the mode.

---

[1] Some people use the less conservative range of -2 to +2 to indicate a normal distribution. You should also note that there is a more sophisticated method for determining whether skewness and kurtosis statistics deviate significantly from 0 (that involves dividing the statistic by its standard error), but this method is often criticized for being biased by the sample size.

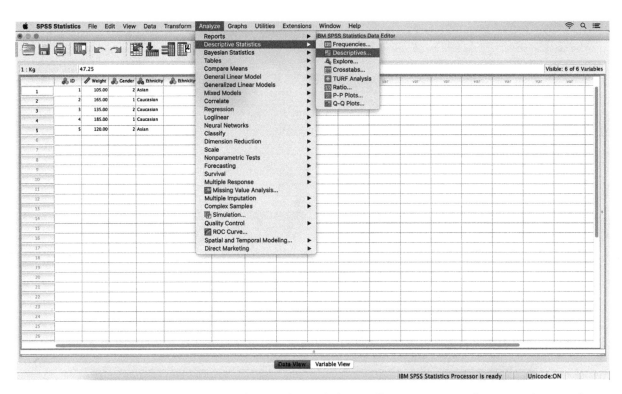

A "Descriptives" dialogue window, like the one shown below, will now appear. Let's practice by transforming the weights into *z* scores. Enter **Weight in Pounds** into the **Variable(s): box** by highlighting it and pressing the **blue arrow**. To obtain the *z* scores for this variable, simply **check the box** next to **Save standardized values as variables**. Press **OK**.

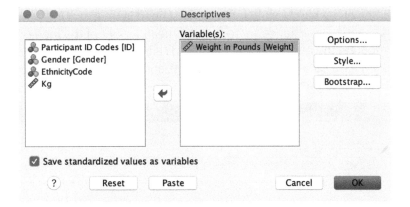

Interpreting the Results

An output window will appear, showing descriptive statistics for the weight variable. Since we are not currently interested in these statistics, you can simply return to the Data View window by clicking on the red star icon that appears in the icon toolbar of the output window (or by closing the output window).

As highlighted in the image shown here, a new column of scores labeled "ZWeight" has been created. This column contains each participant's *z* score on the weight variable. For example, you can see that the participant with the ID code 1 received a *z* score of −1.13, indicating that she has a weight that is 1.13 standard deviations below the mean weight of 142.00 lb. The participant with an ID code of 2 received a *z* score of 0.70, indicating that he has a weight that is 0.70 standard deviations above the mean.

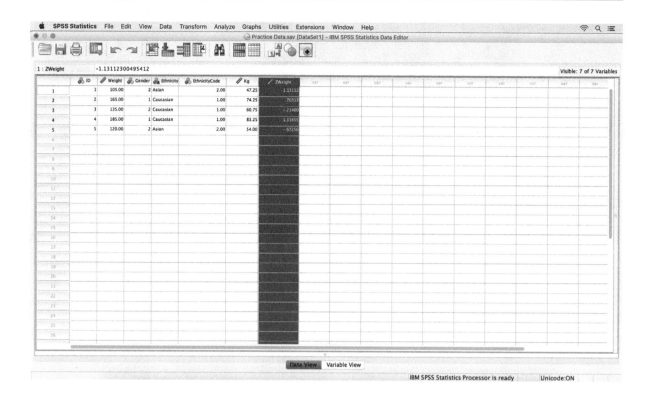

Try On Your Own

Try combining some of the handy tools and tricks described in Chapter One with the analyses described in this chapter. For instance, try to find the mean weight of the men only (hint: select cases before computing the mean weight). Also try to calculate separate z scores for men and women (hint: split the file before creating the z scores), and then use the sort cases function to find the highest z score for each gender.

Correlation

In this chapter, you will learn how to perform, interpret, and report the results of correlation analyses with variables measured on ratio, interval, ordinal, and nominal scales. You will also learn how to conduct, interpret, and report the results of one-tailed correlation and partial correlation analyses.

Learning Objectives

SAMPLE DATA FILES

The SPSS program includes a large number of sample data files. To follow the demonstrations in this chapter, you will need to open a sample data file entitled "Employee data."

Opening Sample Data Files

Start by opening the SPSS program by clicking on the program icon or file name (SPSSStatistics). Select the **Sample Files** option on the "Welcome to IBM SPSS Statistics" window displayed below. Next select **English** and scroll through the sample data files until you find the one entitled "Employee data.sav." Highlight this file and then click **Open.**

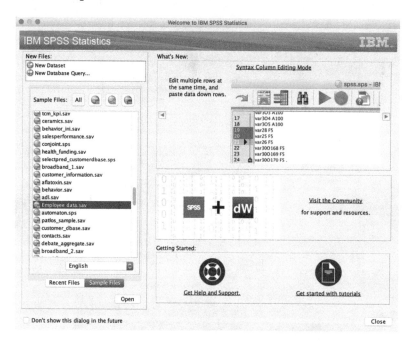

The employee data set contains data on 474 participants ($N = 474$). Whenever you open a data set that you did not create, you should immediately examine the Variable View window to familiarize yourself with the variables, their scales of measure, and values. Take a minute now to go to the **Variable View** window so you can familiarize yourself with this data set. By looking at the variable names and labels, you will see that the file contains data on each participant's gender, birth date, years of education (named "educ"), type of job (named "jobcat"), current salary in dollars (named "salary"), beginning salary in dollars (named "salbegin"), the number of months in the job (named "jobtime"), the number of months of previous experience (named "prevexp"), as well as information on whether the individual is a minority (named "minority"). The Values column shows that individuals working in a clerical position were coded with a 1, those working in a custodial position were coded with a 2, and those working in a managerial position were coded with a 3. Finally, this column shows that minorities were coded with a 1, and non-minorities were coded with a 0.

SCATTERPLOTS

It is appropriate to calculate a correlation coefficient only when the variables you wish to correlate show a linear relationship. This is because the analysis involves the assumption that the relationship between the variables is linear, and the magnitude of the correlation coefficient will be underestimated if the variables show a curvilinear relationship. Thus, before conducting any correlation analysis you should first create a scatterplot to ensure that the variables show a linear relationship.

Generating Scatterplots (Graphs→Legacy Dialogs→Scatter/Dot)

We will begin by creating a scatterplot to examine the relationship between current salary and beginning salary. To create a scatterplot, you will need to use the upper toolbar to go to **Graphs→Legacy Dialogs→Scatter/Dot**.

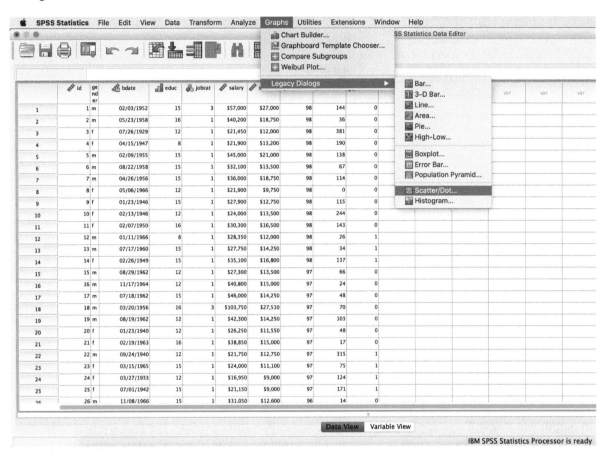

A "Scatter/Dot" dialogue window, like the one displayed below, will appear. Click on the **Simple Scatter** option and then click **Define.**

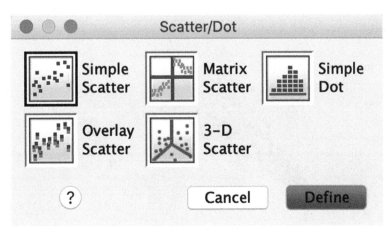

Next, a "Simple Scatterplot" dialogue window, like the one shown below, will appear with all of the variables in the data set listed on the left side. Highlight the variable **Current Salary** by clicking on it, and move it over to the **Y Axis: box** by clicking on the corresponding **blue arrow**. Next, highlight the variable **Beginning Salary** by clicking on it, and move it over to the **X Axis: box** by clicking on the corresponding **blue arrow**. Click **OK** to close the dialogue window.

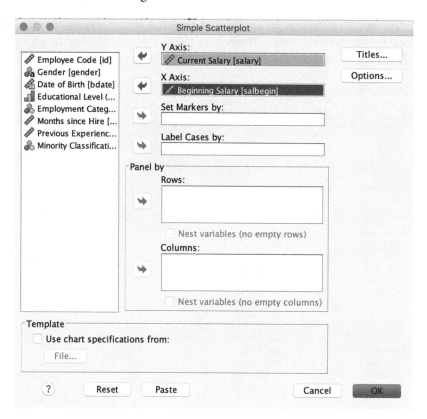

Interpreting Scatterplots

An output window will now appear, displaying the scatterplot shown below. The scatterplot clearly shows that the relationship between current salary and beginning salary is linear (the dots are not falling on a curve). Since the relationship is linear, it is appropriate to calculate the correlation coefficient.

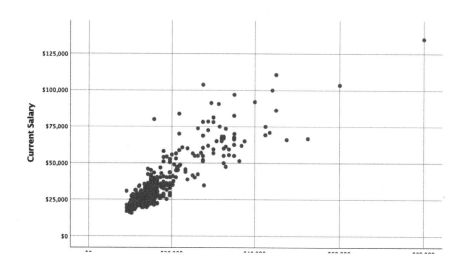

Usually, scatterplots will not look quite as perfect as the one shown previously. When one of the variables is nominal with two levels/values (e.g., gender), scatterplots can be particularly strange looking and difficult to interpret. (The data points will all fall on two separate lines because there are only two levels/values of the nominal variable.) In most cases you should assume the relationship is linear unless the scatterplot depicts a clear curvilinear relationship.

Try On Your Own

Practice creating some scatterplots on your own. Try to create scatterplots depicting the relationships between educational level and current salary as well as between educational level and beginning salary. You will see those relationships are not as clear-cut as the relationship we examined between current salary and beginning salary. Since educational level is an ordinal variable with a limited number of levels, the data points all fall on separate lines. However, since the scatterplots show that there are no clear curvilinear relationships, we will proceed to compute the correlation coefficients.

PEARSON CORRELATION COEFFICIENTS (*r*)

Correlations coefficients are most commonly computed using two variables measured on interval or ratio scales. This type of correlation coefficient is formally referred to as a Pearson correlation coefficient and is symbolized as *r*.

Computing Pearson Correlation Coefficients (Analyze→Correlate→Bivariate)

Let's start by considering the correlation between current salary and beginning salary, each of which is measured on a ratio scale. To compute the Pearson correlation coefficient for these variables, you will need to use the upper toolbar to go to **Analyze→Correlate→Bivariate**.

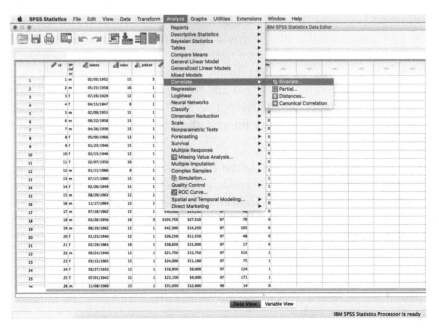

A dialogue window labeled "Bivariate Correlations" will now open with all of the variables in the dataset listed on the left side.[1] Click on **Current Salary,** and use the **blue arrow** to move it into the **Variables: box.** Next, click on **Beginning Salary,** and use the **blue arrow** to move it into the **Variables: box**. Press **OK** to close the dialogue window and initiate the analysis.

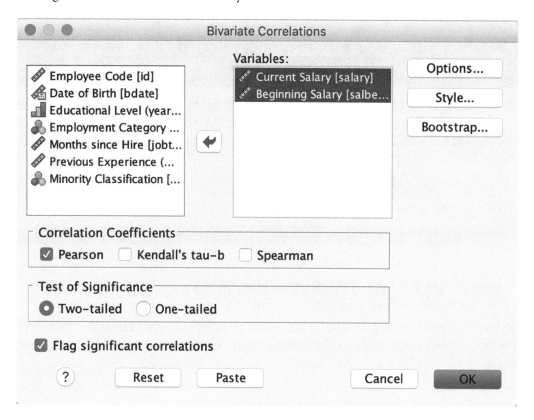

Interpreting the Results

The following "Correlations" table will now appear in the output window. The rows labeled "N," indicate how many pairs of scores (e.g., participants) were included in the analysis. If you were missing the beginning salary of one participant, a value of 473 would appear in these rows instead of 474, because SPSS will include only participants who have data on both of the variables in the analysis. The rows labeled "Pearson Correlation" contain the Pearson correlation coefficients. First, you should see that the correlation between current salary and current salary is 1 and that the correlation between beginning salary and beginning salary is also 1. The correlation between a variable and itself will always equal 1, so this is not terribly interesting or informative. Of primary interest, the table shows that the correlation between current salary and beginning salary is .88.

According to Cohen's (1988) guidelines, correlation coefficients of .10 are generally interpreted as small, correlations of .30 are generally interpreted as moderate, and those of .50 and above are generally interpreted as large. Thus, our correlation of .88 indicates that there is a large positive correlation between beginning salary and current salary. The fact that the coefficient is positive indicates that as beginning salary increases, so does current salary. If the coefficient were negative, it would indicate that higher beginning salaries are associated with lower current salaries.

---

[1] Since the gender variable is a string variable (i.e., a variable that contains text), it is not contained in the list. Remember that SPSS will not analyze variables entered using text.

Correlations

		Current Salary	Beginning Salary
Current Salary	Pearson Correlation	1	.880[**]
	Sig. (2-tailed)		.000
	N	474	474
Beginning Salary	Pearson Correlation	.880[**]	1
	Sig. (2-tailed)	.000	
	N	474	474

**. Correlation is significant at the 0.01 level (2-tailed).

Determining Statistical Significance

In addition to knowing the size and direction of the correlation between two variables, it is also useful to know whether the correlation coefficient is statistically significant. Statistically significant effects are those that have a low probability of occurring due to chance alone. Specifically, an effect is considered statistically significant if there is a low probability that it would be obtained if there is no real effect in the population. In the context of correlation, a correlation coefficient is considered statistically significant if there is a low probability that the correlation would be found if the real correlation in the population was actually 0.

Typically, to be considered statistically significant, the probability that an effect is due to chance alone must be .05 (i.e., 5%) or less. Sometimes, however, researchers set a more stringent criterion and require that this probability be .01 (i.e., 1%) or less before they will consider the effect statistically significant. This criterion or threshold is referred to as the alpha level (symbolized α), and it must be decided on before conducting the study. Once again, alpha is a threshold that we set before conducting a study for our willingness to conclude that there is an effect (e.g., that there is a correlation) when in reality there is no effect in the population and our results are just due to chance. And this threshold is always set low (typically, either at .05 or .01).

Information about whether a correlation coefficient is statistically significant can be found in the rows labeled "Sig. (2-tailed)." The *p* values or significance levels provided in these rows reflect the probability that we would obtain our result, or a result more

© zetwe/Shutterstock.com

extreme, if there is no real effect in the population. In other words, it is the probability that chance alone is operating and, in reality, the correlation in the population is 0. If the p value listed in the "Sig. (2-tailed)" row is less than or equal to the alpha level set prior to conducting the study, the correlation is considered statistically significant. If the p value in the "Sig. (2-tailed)" row is greater than the alpha level set, the correlation is not statistically significant. To use the analogy of the limbo dance, you can think of the alpha level as a limbo bar (it is the threshold we set) and the p value as the person (it is the value we want to be under alpha). When p is under alpha, our effect is statistically significant (and it's time to party)!

As already mentioned, the alpha level is typically set at .05, and as such, we will use that alpha level. (Again, this decision really should have been made before we conducted the study and ran the analysis, but at that point I hadn't introduced the concept to you.) The "Correlations" table displays the p value for the correlation between current salary and beginning salary as .000. This value indicates that there is less than a .001 chance that this correlation, or one larger, would be found if only chance is operating and, in reality, there is no correlation between these variables. Since this p value is less than .05 (the alpha level we set), we can conclude that there is a statistically significant correlation between the variables. The stars next to the correlation coefficients in the table also indicate that the correlation is significant using the standard .05 alpha level. If stars were not present, it would indicate that the correlation is not statistically significant using the standard .05 alpha level.

Reporting the Results

The results of correlation analyses are reported using the following format: $r(\text{df}) = .\#\#, p = .\#\#$. Note that the r and p values do *not* have leading 0s before the decimal point because these values can never exceed a value of 1. (See the sections on Reporting Decimal Remainders and Rounding in Chapter One for a complete description of reporting decimal remainders.) Also recall that American Psychological Association (APA) style requires us to italicize all statistical notation (except Greek characters), so the symbols r and p should both be italicized.

We have already established that r is used to symbolize the Pearson correlation coefficient and p symbolizes the p value or significance level. Now all you need to know is that df stands for degrees of freedom. The number of degrees of freedom is not provided in the output table, but degrees of freedom for correlations are very easy to calculate; they simply equal the size of the sample minus 2 ($df = N - 2$). As described earlier, the sample size is 474, so our degrees of freedom are: $474 - 2 = 472$. Degrees of freedom are discrete numbers, so no decimal remainder should be reported. Therefore, for the results obtained previously we could report the following:

There is a large, positive correlation between current salary and beginning salary that is statistically significant, $r(472) = .88, p < .001$.

Try On Your Own

Try to compute the Pearson correlation between current salary and months since hire. Use the conventional alpha level of .05, and practice reporting the results using APA style.

SPEARMAN RANK ORDER CORRELATION COEFFICIENTS (r_s)

Correlation coefficients can also be computed with variables measured on ordinal scales. When one of the variables is measured on an ordinal scale and the other variable is measured on an ordinal, interval, or ratio scale, the correlation is referred to as a Spearman rank order correlation coefficient, or Spearman's rho (pronounced row), and it is symbolized as r_s.

Computing Spearman Rank Order Correlation Coefficients (Analyze→Correlate→Bivariate)

Let's begin by computing some Spearman rank order correlations using the variable educational level, which has been defined as an ordinal variable in this data set[2]. We will correlate educational level with both beginning salary and current salary. We will stick with the convention and set alpha at .05. Go to **Analyze→Correlate→Bivariate**. Put **Current Salary**, **Beginning Salary**, and **Educational Level** in the **Variables: box** by highlighting each and clicking on the **blue arrow**. Do **NOT** check the box labeled 'Spearman' (see the following "Important Note" for an explanation). Click **OK** to initiate the analyses.

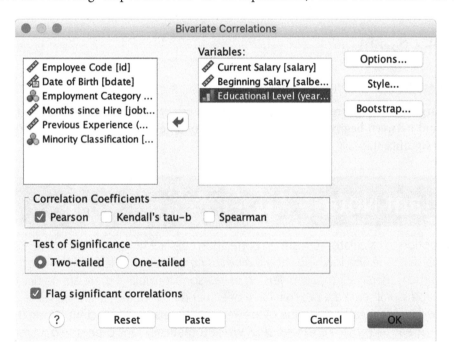

Interpreting the Results

A "Correlations" table will now appear in the output window. It shows all of the correlations between the three variables (in the rows labeled "Pearson Correlation") as well as the associated p values [in the rows labeled "Sig. (2-tailed)"], and sample sizes (in the rows labeled "N"). Specifically, the bottom portion of the table shows that there are large positive correlations between educational level and current salary (.66) and educational level and beginning salary (.63) that are statistically significant ($p < .001$). Remember that we

---

[2] Frankly, I entirely disagree with this classification of years of education as an ordinal variable. I think it should be coded as a Scale variable because the levels of the variable represent number of years of education, but I will go along with it for the pedagogical purpose of illustrating a Spearman rank order correlation.

will need to compute the degrees of freedom ourselves by subtracting 2 from N. Since $N = 474$ for both correlations, our degrees of freedom are $474 - 2 = 472$.

Correlations

		Current Salary	Beginning Salary	Educational Level (years)
Current Salary	Pearson Correlation	1	.880**	.661**
	Sig. (2–tailed)		.000	.000
	N	474	474	474
Beginning Salary	Pearson Correlation	.880**	1	.633**
	Sig. (2–tailed)	.000		.000
	N	474	474	474
Educational Level (years)	Pearson Correlation	.661**	.633**	1
	Sig. (2–tailed)	.000	.000	
	N	474	474	474

**. Correlation is significant at the 0.01 level (2–tailed).

Reporting the Results

Based on these results, we can report the following:

> There are large, positive correlations between current salary and educational level, $r_s(472) = .66$, $p < .001$, and between beginning salary and educational level, $r_s(472) = .63$, $p < .001$, that are statistically significant.

Important Note

While the option to calculate a Spearman correlation coefficient is provided in the "Bivariate Correlations" dialogue window, you should *avoid using this option*. The formula for calculating Spearman's rho is merely a simplified version of Pearson's formula. If there are no ties on the ordinal variables, or if there are ties that have been handled properly, then the value of Spearman's rho will be identical to the value of the Pearson correlation coefficient. When the option to calculate Spearman's rho is checked, SPSS will automatically rank order the data for both of the variables (rather than just the variable defined as ordinal). For this reason, you should never use the option to compute a Spearman correlation coefficient unless both of the variables you are correlating were measured on ordinal scales. Even in this case the value of the Pearson correlation will be identical to the value of Spearman's rho (if ties are properly handled*), so many researchers will still calculate the Pearson correlation between two ordinal variables.

*Ties on ordinal variables should be handled in a specific way, namely, by assigning the mean rank of the tied variables (e.g., if there is a tie for the rank of 2, both rankings should be changed to 2.5, and the ranking of 3 should be skipped). Ties in the educational level variable have not been properly handled in this data set. The option of properly rank ordering the data is available in SPSS (Transform→Rank Cases). Since the mishandling of ties has a very small effect on the magnitude of the correlation coefficients (properly ranking the data increases each correlation by less than .02), we will not worry about ranking the data.

POINT BISERIAL CORRELATION COEFFICIENTS (r_{pb})

Correlations can also be computed with variables measured on nominal scales as long as the nominal variables are dichotomous (they contain only two categories). Correlations should never be computed with variables measured on nominal scales that contain more than two categories because the results would be meaningless (e.g., the correlation between undergraduate major and hair color could not be interpreted in any meaningful way). When one variable is measured on a nominal scale and the other variable is measured on an interval or ratio scale, the correlation is referred to as a point biserial correlation, and it is symbolized as r_{pb}.

Computing Point Biserial Correlation Coefficients (Analyze→Correlate→Bivariate)

We will practice computing point biserial correlations by correlating gender with both current salary and beginning salary. We will stick with the convention and set alpha at .05. Since gender is entered as a string (text) variable, we will not be able to perform any analyses with it until it has been recoded using a numeric code. Go to **Transform→Recode into Different Variables.** Label the recoded variable **Gender-Code,** and then proceed to recode the current gender code "**m**" to "**1**" and "**f**" to "**2.**" (If you forget how to do this, refer to the section entitled "Recoding Variables" in Chapter One for a complete demonstration.) Save this dataset as we will use this file with the numerically recoded gender variable again in subsequent chapters.

Now that we have a numerically coded gender variable, we can compute correlations with it. Point biserial correlations and phi coefficients are computed in the same way as Pearson correlation coefficients,[3] so using the upper toolbar, simply go to **Analyze→Correlate→Bivariate.** Enter **Current Salary**, **Beginning Salary,** and **GenderCode** in the **Variables** box by highlighting each and clicking on the **blue arrow.** Click **OK** to initiate the analyses.

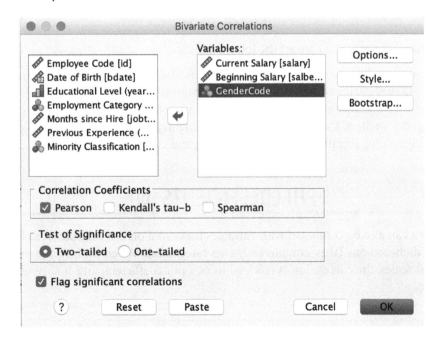

---

[3] While there are different formulas for calculating these coefficients, they are merely simplified versions of Pearson's formula, and they produce the same results. Since SPSS is doing all of our work, we don't need to worry about simplified formulas!

Interpreting the Results

The following "Correlations" table will now appear in the output window. The bottom portion of the table shows that there are moderately sized negative correlations between gender and current salary ($-.45$) and gender and beginning salary ($-.46$) that are statistically significant ($p < .001$). Before we can fully interpret these correlations, we need to consider the code used for the gender variable. We assigned men a code of 1 (a lower value) and women a code of 2 (a higher value). The negative coefficients indicate that high values on one variable are associated with low values on the other. As such, we can conclude that being a woman (a higher value) is associated with earning significantly lower beginning salaries as well as with significantly lower current salaries. Similarly, being a man (a lower value) is associated with earning significantly higher beginning and current salaries.

Correlations

		Current Salary	Beginning Salary	GenderCode
Current Salary	Pearson Correlation	1	.880**	−.450**
	Sig. (2−tailed)		.000	.000
	N	474	474	474
Beginning Salary	Pearson Correlation	.880**	1	−.457**
	Sig. (2−tailed)	.000		.000
	N	474	474	474
GenderCode	Pearson Correlation	−.450**	−.457**	1
	Sig. (2−tailed)	.000	.000	
	N	474	474	474

**. Correlation is significant at the 0.01 level (2−tailed).

Reporting the Results

On the basis of these results we can report the following:

There are moderately sized, negative correlations between gender and current salary, $r_{pb}(472) = -.45$, $p < .001$, and between gender and beginning salary, $r_{pb}(472) = -.46$, $p < .001$, that are statistically significant. Since a code of 1 was used to designate men and a code of 2 was used to designate women, the negative correlation coefficients indicate that being a woman is associated with earning significantly lower beginning and current salaries.

PHI COEFFICIENTS (Φ)

Again, correlations can also be computed with variables measured on nominal scales as long as the nominal variables are dichotomous (they contain only two categories). When both of the variables are measured on nominal scales, the correlation is referred to as a phi coefficient, and it is symbolized using the Greek character ϕ.

Computing Phi Coefficients (Analyze→Caorrelate→Bivariate)

Next, we will compute the phi coefficient for the relationship between minority classification and gender. We will set alpha at .05. Note that we can only use minority classification as a variable in our analysis because it has been coded in a binary manner and therefore only contains two categories (minority and nonminority). Remember that meaningful correlations cannot be computed with nominal variables that contain more than two categories. To compute the phi coefficient, you will once again need to go to **Analyze→Correlate→Bivariate.** Remove the variables (Current Salary, Beginning Salary) you put in the Variables: box for the previous analysis, by clicking on them and using the blue arrow to move them back over to the list of variables on the left. Then, enter **Minority Classification** into the **Variables: box** by highlighting it and clicking on the **blue arrow**. As shown below, both GenderCode and Minority Classification should now appear in the Variables: box. Click **OK** to initiate the analyses.

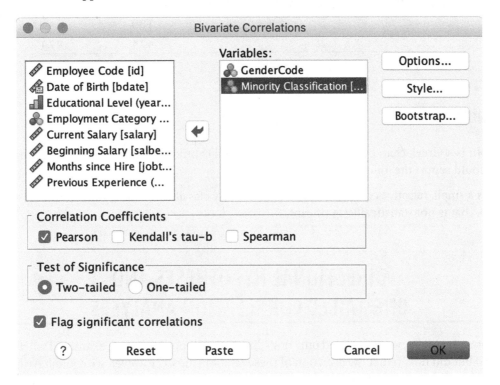

Interpreting the Results

The following "Correlations" table will now appear in the output window. Since the *p* value displayed in the table is higher than our alpha level of .05, we must conclude that the correlation is not statistically significant and therefore that there is no correlation between the variables. Nevertheless, it is good to practice interpreting the direction of phi coefficients. Before you can interpret any phi coefficient, you need to consider the codes used for both variables. As we just reviewed, we assigned men a code of 1 (a lower value) and women a code of 2 (a higher value). By looking in the Values column in the Variable View window, you will find that individuals who are not minorities were labeled with a 0 (a lower value), while those who are minorities were labeled with a 1 (a higher value).[4] The presence of a negative coefficient indicates that higher values on one variable tend to be associated with lower values on the other variable. Thus, it appears that being a woman is associated with not being a minority and that being a man is associated with being a minority. Once again, however, this relationship is not statistically significant (or terribly interesting)!

---

[4] You may also notice that a code of 9 was used for missing data. Since missing data are not actually considered a level of the variable (the code is not treated as an actual quantitative value), it is a true dichotomous variable and can be analyzed using correlation.

Correlations

		GenderCode	Minority Classification
GenderCode	Pearson Correlation	1	−.076
	Sig. (2–tailed)		.100
	N	474	474
Minority Classification	Pearson Correlation	−.076	1
	Sig. (2–tailed)	.100	
	N	474	474

Reporting the Results

Note that phi is a Greek character and as such, the symbol should not be italicized. On the basis of these results we could report the following:

There is a small, negative correlation between minority classification and gender, $\phi(472) = -.08$, $p = .10$, that is not statistically significant.

DIRECTIONAL HYPOTHESES AND ONE-TAILED CORRELATION ANALYSES

In the previous examples, we considered only non-directional hypotheses. (We examined whether correlations existed but did not predict the direction of these relationships.) However, we could have instead used directional hypotheses, in which we predicted the direction of the relationships. When our hypothesis is non-directional, we use a two-tailed analysis (we consider two possible outcomes, a positive relationship *and* a negative relationship). In contrast, when our hypothesis is directional, we use a one-tailed analysis (considering only one possible outcome, a positive *or* a negative relationship).

In the "Try on Your Own" section that followed the section on "Pearson Correlation Coefficients," you were asked to practice computing the Pearson correlation between current salary and months since hire. You should have found that the correlation between these variables is not statistically significant, $r(472) = .08$, $p = .07$. Since there is no reason to believe that less experience (fewer months since hire) would be associated with higher current salaries, we could have instead made a directional hypothesis. Specifically, we could have made the directional hypothesis that there is a positive correlation between current salary and the number of months since hire. Let's assume we did that. Once again, we will stick with convention and set alpha at .05 (one-tailed).

Conducting One-Tailed Correlation Analyses (Analyze→Correlate→Bivariate)

In order to conduct a one-tailed correlation analysis, you once again need to go to **Analyze→ Correlate→Bivariate.** Using the "Bivariate Correlations" dialogue window, remove GenderCode and Minority Classification from the Variables: box. Then move the variables **Current Salary** and **Months since Hire** into the **Variables box** using the **blue arrow**. To change the default from a two-tailed analysis to a one-tailed analysis, simply click on the **One-tailed** option in the section of the dialogue window labeled "Test of Significance." Finally, click **OK** to execute the analysis.

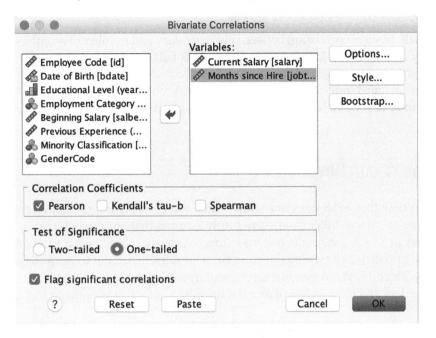

Interpreting the Results

The following "Correlations" table will now appear in the output window. As you can see, the value of the correlation coefficient does not change; it is still $r(472) = .08$. The only value that changes with a one-tailed analysis is the p value. First, you should see that the rows containing the p values are now labeled "Sig. (1-tailed)" to indicate that they are p values for a one-tailed test. The p value for a one-tailed test will always be exactly half of the value it would be for a two-tailed test. The p value for the two-tailed correlation between these variables was .067. The p value for this one-tailed correlation is now half that size; it is .034. Since .03 is less than our alpha of .05, this one-tailed correlation is statistically significant! But wait; before making any conclusions, we need to check to make sure that the correlation is in the same direction as the one we predicted. It is in the positive direction, so we are ready to report this result.

Correlations

		Current Salary	Months since Hire
Current Salary	Pearson Correlation	1	.084[*]
	Sig. (1–tailed)		.034
	N	474	474
Months since Hire	Pearson Correlation	.084[*]	1
	Sig. (1–tailed)	.034	
	N	474	474

*. Correlation is significant at the 0.05 level (1–tailed).

Reporting the Results

Using APA style, we can now report the following:

> There is a small, positive correlation between current salary and number of months since hire that is statistically significant, $r(472) = .08$, $p = .03$ (one-tailed).

Note that we have included "one-tailed" in parentheses next to the p value to indicate that our hypothesis was directional, and our p value is one-tailed. It is important to do this whenever reporting the results of a one-tailed test.

Predicting the Wrong Direction

It is important to note that SPSS does not know which direction we are predicting when we select the option to run a one-tailed test. SPSS always generously assumes that our prediction is in the correct direction (the direction in which the results end up falling). As such, it is important to consider how things would change if we predicted the wrong direction (if the results turned out to be in the opposite direction to the one we predicted). Let's assume our directional hypothesis was that there is a *negative* correlation between current salary and the number of months since hire (a foolish hypothesis indeed).

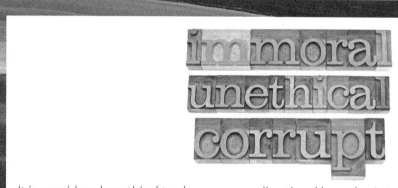

It is considered unethical to change a non-directional hypothesis to a directional hypothesis after discovering that the effect is significant for a one-tailed test but not a two-tailed test. It would be comparable to changing alpha from .05 to .10 after discovering that your p value is .09. It is even worse to change the direction of your hypothesis after discovering the results are in the opposite direction to the one you predicted!

If we were to run the one-tailed correlation analysis, we would get the same "Correlations" table shown previously, indicating a positive correlation of .08 and a p value of .03. However, since the results turned out to be in the opposite direction to the one we predicted (in this scenario, we are predicting a negative correlation, but the results show a positive correlation), we would need to adjust the p value by subtracting it from 1. So, the p value that we would report in this scenario would actually be .97 ($1 - .0337 = .97$). Thus, our reported results would be: $r(472) = .08$, $p = .97$ (one-tailed). Since in this scenario, $p > .05$, we would need to conclude that there is not a significant negative correlation between current salary and months since hire. Note: in this scenario, we *cannot* conclude that there is a significant positive correlation between the variables (see box above). This is the risk we take when we make a directional hypothesis. If the results turn out to be significant in the opposite direction, we must adjust the p value and conclude the result is not significant. However, with this risk comes greater power to find an effect in the predicted direction (because our p values are cut in half with one-tailed tests).

PARTIAL CORRELATION (*pr*)

Correlation does not permit determination of causation. The third variable problem is one reason why we can never infer causation based on the results of a correlational study. In other words, a correlation between two variables may appear simply because both of the variables are related to some extraneous third variable. For instance, if we assessed 100 participants and discovered that the correlation between their depression levels and the number of fast food meals they consumed last month was, $r(98) = .35, p < .001$, we could not conclude that consuming more fast food causes people to become more depressed or that higher levels of depression cause people to consume more fast food, because a third variable may be responsible for the relationship. In this case, higher fast food consumption may be associated with reduced income, and reduced income may be associated with higher levels of depression, rather than there existing a direct causal relationship between fast food consumption and depression.

The only way to determine causality is through the use of the experimental method because potential confounding (i.e., third) variables are physically controlled. While we cannot physically control for third variables using correlation, we can statistically control for them. We can rule out third variables using a technique called partial correlation. This technique allows us to examine the relationship between two variables (e.g., fast food consumption and depression) after statistically controlling for a potential third variable (e.g., income).[5] The symbol provided in the APA style guide for the partial correlation coefficient is *pr*.

Interpreting Partial Correlations

Interpreting a partial correlation coefficient simply involves comparing the magnitude of the original correlation coefficient with the magnitude of the partial correlation coefficient. The original correlation coefficient is often referred to as the bivariate correlation coefficient. ("Bi" indicates two and "variate" indicates variables, so the bivariate correlation is the relationship between two variables.) When the bivariate correlation coefficient is large and statistically significant, and the partial correlation coefficient is substantially lower and is not statistically significant, it suggests that the variable that was statistically partialed out is a third variable that is responsible for the bivariate correlation. To illustrate, assume you ran a correlation analysis and found that the correlation between depression and fast food consumption was, $r(98) = .35$, $p < .001$. Now assume you ran a partial correlation analysis—correlating depression with fast food consumption, after controlling for income—and found that the partial correlation was, $pr(97) = .09, p = .37$. The difference between these coefficients and the drop from statistical significance suggests that income is responsible for the correlation between depression and fast food consumption. In other words, this outcome suggests that there is little to no relationship between depression and fast food consumption, when income is statistically controlled.

When the bivariate correlation coefficient and partial correlation coefficient are identical or very similar, it suggests that the variable that was partialed out is not a third variable, that it is having little influence on the bivariate correlation. To illustrate, if the bivariate correlation between depression and fast food

---

[5] Note that even if we use partial correlation to rule out third variables, we still cannot determine causation. This is because other third variables may still be at play and because of the directionality problem (i.e., we can't determine whether depression is causing fast food consumption or fast food consumption is causing depression using the correlational method).

consumption was, $r(98) = .35$, $p < .001$, and the partial correlation was, $pr(97) = .32$, $p = .002$, then we could conclude that income does not account for the correlation between depression and fast food consumption. This is because controlling for income had very little effect on the magnitude or significance of the correlation.

When the partial correlation coefficient is statistically significant but substantially lower than the bivariate correlation coefficient, it suggests that the variable that was partialed out is a third variable but also that there is a relationship between the variables of interest that is independent of the third variable. In other words, if a significant relationship still exists between the variables after the potential third variable is controlled, it suggests that a relationship exists between the variables of interest that is independent of the variable that was controlled. However, because controlling for the potential third variable reduced the magnitude of their correlation, we can infer that the third variable was increasing the magnitude of the bivariate correlation. For example, if the bivariate correlation between depression and fast food consumption was, $r(98) = .35$, $p < .001$, and the partial correlation was, $pr(97) = .20$, $p = .05$, you would know that income accounts for *some* of the relationship between depression and fast food consumption (because controlling for income decreased the correlation coefficient by .15) but that there is still a relationship between the depression and fast food consumption that is independent of income (the partial correlation is still statistically significant).

Computing Partial Correlations (Analyze→Correlate→Partial)

Let's try to calculate the partial correlation between current salary and educational level, controlling for beginning salary.[6] Go to **Analyze→Correlate→Partial.**

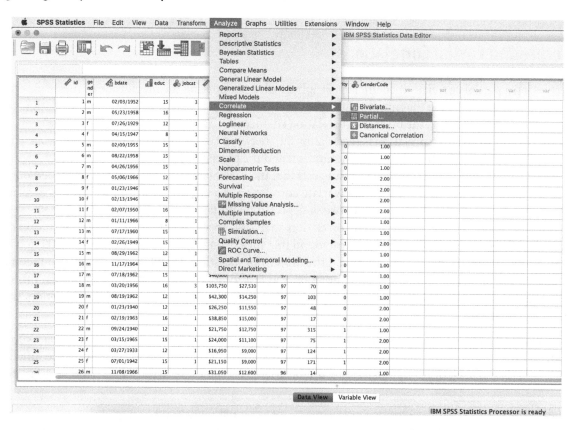

---

[6] Technically, we shouldn't be using an ordinal variable in a partial correlation analysis, but, once again, I completely disagree with the categorization of educational level (in years) as ordinal, so we're going to go ahead and do it anyway.

Move **Current Salary** and **Educational level** into the **Variables box** using the corresponding **blue arrow,** and move **Beginning Salary** in the **Controlling for box** using the corresponding **blue arrow**.

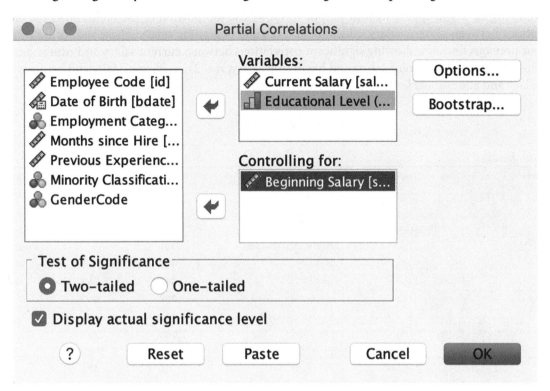

Next click on the **Options tab** in the "Partial Correlations" dialogue window. This will open the "Partial Correlations: Options" dialogue window shown below. **Check** the **Zero-order correlations box**. By checking this option, the output window will contain both the partial correlation and the bivariate correlations (which SPSS refers to as zero-order correlations), allowing us to more easily compare their values. Click **Continue** and then **OK** to close the dialogue windows and initiate the analysis.

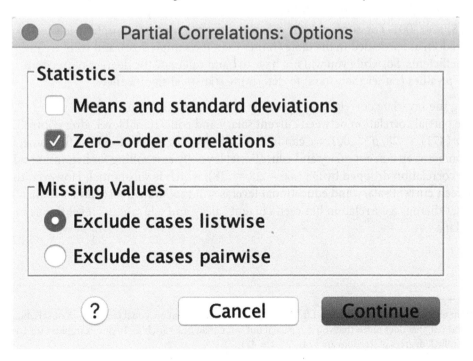

Interpreting the Results

The following table will appear in your output window. The top portion of the table shows the bivariate correlations between current salary, educational level, and beginning salary. These results are consistent with our previous findings, showing significant correlations between current salary and educational level, $r_s(472) = .66, p < .001$, current salary and beginning salary, $r(472) = .88, p < .001$, and between beginning salary and educational level, $r_s(472) = .63, p < .001$.

Correlations

Control Variables			Current Salary	Educational Level (years)	Beginning Salary
−none−[a]	Current Salary	Correlation	1.000	.661	.880
		Significance (2-tailed)	.	.000	.000
		df	0	472	472
	Educational Level (years)	Correlation	.661	1.000	.633
		Significance (2-tailed)	.000	.	.000
		df	472	0	472
	Beginning Salary	Correlation	.880	.633	1.000
		Significance (2-tailed)	.000	.000	.
		df	472	472	0
Beginning Salary	Current Salary	Correlation	1.000	.281	
		Significance (2-tailed)	.	.000	
		df	0	471	
	Educational Level (years)	Correlation	.281	1.000	
		Significance (2-tailed)	.000	.	
		df	471	0	

a. Cells contain zero-order (Pearson) correlations.

The bottom section of the table shows the partial correlation between current salary and educational level, controlling for beginning salary. As shown in the table, the partial correlation is statistically significant, $pr(471) = .28, p < .001$. For some reason, when the partial correlation option is used to analyze data, degrees of freedom are provided in the table (rather than the sample size), and stars are not placed next to significant coefficients. So, while you will not have to hand calculate the degrees of freedom,[7] you will have to rely on the p values (rather than stars) to determine statistical significance.

By comparing the bivariate correlation between current salary and educational level, $r_s(472) = .66, p < .001$, with the partial correlation between current salary and educational level, after controlling for beginning salary, $pr(471) = .28, p < .001$, we can infer that beginning salary accounts for some, but not all, of the correlation between current salary and educational level. By controlling for beginning salary, the magnitude of the correlation dropped by .38 (.66 − .28 = .38), which is substantial. However, the partial correlation between current salary and educational level is still moderate in size and statistically significant, suggesting that there is a correlation between current salary and educational level that is independent of beginning salary.

---

[7] For those of you who are interested, degrees of freedom for partial correlation are equal to $N - 2 - $ # of variables being partialed out (and yes, that does mean that you can partial out more than one variable). In our example, with 1 variable being statistically controlled, degrees of freedom are 474 − 2 − 1 = 471.

Reporting the Results

The following results could be reported:

The partial correlation between current salary and educational level, after controlling for beginning salary, is statistically significant, $pr(471) = .28$, $p < .001$, however it is substantially lower than the bivariate correlation between these variables, $r_s(472) = .66$, $p < .001$. This suggests that much of the correlation between current salary and educational level is simply due to beginning salary. However, there also appears to be a relationship between current salary and educational level that is independent of beginning salary.

Try On Your Own

Try to compute the partial correlation between beginning salary and current salary, after controlling for educational level. Interpret the results and then report them using APA style.

Simple Regression

Learning Objectives

In this chapter, you will learn how to conduct a simple regression analysis. You will learn how to determine whether a regression model, and the individual predictor contained in the model, are statistically significant. You will also learn how to interpret the multiple correlation, coefficient of determination, and standard error of estimate. Finally, you will learn how to use the results of a simple regression analysis to construct the least-squares regression line and make predictions.

INTRODUCTION TO REGRESSION

Regression is simply an extension of correlation. We use correlation when we want to determine the strength and direction of relationship between variables, and we use regression when we want to use the information about those relationships to make predictions. We will begin with a simple regression analysis, which is a regression analysis that involves the use of a single predictor variable.

Let's start by examining whether we can use number of years of education to predict beginning salary. Thus, for this analysis beginning salary will be our criterion variable (i.e., the variable we want to predict or the Y variable) and years of education will be our predictor variable (i.e., the variable we will use to make our predictions or the X variable).

We will once again use the sample data set "Employee data" for the demonstrations in this chapter. If you saved the file used for the demonstrations in Chapter Three, you should use it. If you didn't save the file, the original Employee data file can be found in your SPSS sample files (refer to the "Opening Sample Data Files" section of Chapter Three if you forget how to locate and open these files).

Conducting a Simple Regression Analysis (Analyze→Regression→Linear)

To conduct the analysis, use the upper toolbar to go to **Analyze→Regression→Linear**.

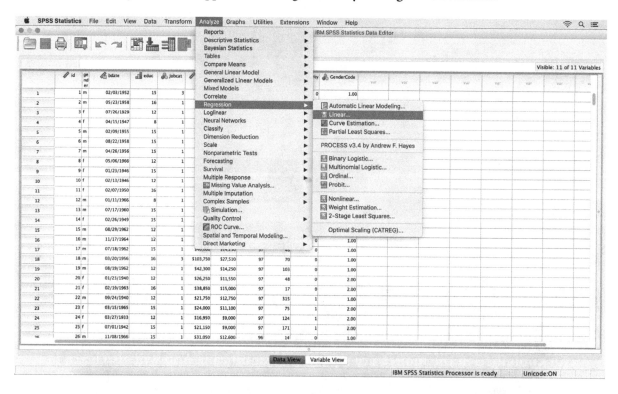

A "Linear Regression" dialogue window, like the following, will now open. Enter **Beginning Salary** (the criterion variable) into the **Dependent: box** using the corresponding **blue arrow**. Next enter **Educational level** (the predictor variable) into the **Independent(s): box** using the corresponding **blue arrow**. Click **OK**.

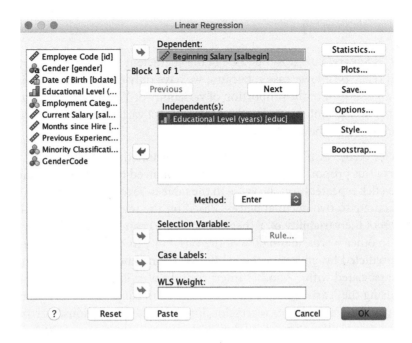

Interpreting the Results
Assessing the Model's Accuracy

A number of tables will now appear in the output window. We will start by reviewing the "Model Summary" table shown here.

Model Summary

Model	R	R Square	Adjusted R Square	Std. Error of the Estimate
1	.633[a]	.401	.400	$6,098.259

a. Predictors: (Constant), Educational Level (years)

The Correlation Coefficient (r)

The column labeled "R" displays the size of the correlation between the variables. In the context of regression, this value can also be interpreted as the correlation between actual Y scores and what we would predict them to be using the regression equation ($r_{YY'}$). The value displayed in this table will always be positive, so the direction of relationship cannot be inferred using this value. Since we are using simple rather than multiple regression (i.e., we only have one predictor variable), the capitalized R shown in the table should really be a lowercase r, but sometimes SPSS doesn't make these subtle distinctions. Note that the higher the correlation coefficient, the more accurate our predictions will be. A correlation of .63 is quite large, so we can expect our predictions of people's beginning salaries based on their years of education to be quite accurate.

The Coefficient of Determination (r^2)

The column labeled "R Square" contains the value of the coefficient of determination (r^2). Once again, the capitalized R should really be a lowercase r, because we have only one predictor variable. The coefficient of determination is an indicator of the proportion of variability of Y that can be accounted for by X. In the context of regression, it is often interpreted as the proportion of variability of Y that can be predicted by X. It is considered an indicator of the effect size.

The "Model Summary" table displays a coefficient of determination of .40, which we would report as $r^2 = .40$. To transform this proportion into a percentage, you need to multiply the value by 100. However, you will first need to click repeatedly on the value in the table to reveal the full decimal remainder in order to keep the value accurate to two decimal places. Based on the revealed value (of .400937), we can now conclude that 40.09% of the variability of Y (beginning salary) can be accounted for, and predicted by, X (educational level). In other words, about 60% of the variability in beginning salary is still not accounted for, and cannot be predicted by, educational level (i.e., we don't know what that remaining variability in beginning salary is associated with). Another interpretation of r^2 is that it reflects the proportion reduction in error from using the least-squares regression line to make predictions rather than the mean. In other words, if we use the least-squares regression line to make predictions, our predictions will have 40.09% less error, on average, than they would if we just predicted the mean of the Y variable for everyone (we predicted everyone would have the average beginning salary).

The table also displays an "Adjusted R Square" value. For simple regression, the adjusted r^2 will always be pretty much identical to r^2. You will notice bigger differences between the adjusted R^2 and R^2 as you add more predictor variables in multiple regression analyses (which are covered in Chapter Five). While R^2 will almost always increase as you add more predictor variables (as long as the predictor variables are correlated with the criterion variable), adjusted R^2 adjusts for the number of predictors used, and its value increases only if a predictor improves the prediction more than would be expected based on chance.

The Standard Error of Estimate (*SEE*)

The standard error of estimate, which is abbreviated *SEE*, is an indicator of the amount of error our predictions will contain. As such, this value allows us to determine how much confidence we should have in our predictions. Specifically, the standard error of estimate reflects the average distance between the actual Y scores in the data set and what they would be predicted to be using the regression equation. Therefore, it provides an indicator of the average amount of error that we can expect in our predictions.

The value of the standard error of estimate can be found in the column labeled "Std. Error of the Estimate" in the "Model Summary" table. A quick glance at this table reveals that *SEE* = $6,098.26. This value means that if we use the regression equation to predict people's beginning salaries based on their level of education, then, on average, we can expect our predictions to be off by $6,098.26.

Assessing Statistical Significance

The Regression Model

The determination of whether the regression model is statistically significant involves assessing whether the coefficient of determination (r^2) is significant. In other words, it involves assessing whether we can predict a significant proportion of variability in Y (the criterion), using X (the predictor). If a regression model is statistically significant, it suggests that we can reliability predict the criterion using the predictor.

Information on whether r^2, and therefore the regression model, is statistically significant is contained in the following "ANOVA" table.

ANOVA[a]

Model		Sum of Squares	df	Mean Square	F	Sig.
1	Regression	1.175E+10	1	1.175E+10	315.897	.000[b]
	Residual	1.755E+10	472	37188762.8		
	Total	2.930E+10	473			

a. Dependent Variable: Beginning Salary

b. Predictors: (Constant), Educational Level (years)

In the context of regression, the F statistic is used to determine the significance of the regression model. As shown in the column labeled "F," the value of the F statistic for this regression model is 315.90. The p value provided under the column labeled "Sig." is less than .001. Since this value is less than .05, this regression model would be considered statistically significant. The table also shows the values of the degrees of freedom, in the column labeled "df." F statistics are always associated with two degrees of freedom values. The first degree of freedom value that we will need to report is listed in the row labeled "Regression," and the second degree of freedom value that we will need to report is listed in the row labeled "Residual." These values are reported in parentheses and are separated by a comma. So they would be reported as (1, 472). Since the assessment of the significance of the regression model involves evaluating whether the coefficient of determination is statistically significant, it is customary to report the coefficient of determination along with the F statistic, degrees of freedom, and p value. To illustrate, using APA style we would report that the regression model is statistically significant, $r^2 = .40$, $F(1, 472) = 315.90$, $p < .001$.

The Predictor

In addition to determining whether the regression model is statistically significant, it is typically important to assess whether the individual predictors are statistically significant. Since simple regression involves only one predictor variable, if the overall regression model is significant, then the predictor will be always be significant and vice versa. As you will see in Chapter Five, this is not always the case for multiple regression, which involves more than one predictor variable.

Information about the predictor variable can be found in the "Coefficients" table. First, the table displays the unstandardized and standardized slopes for the relationship between the predictor and criterion variables. The unstandardized slope is displayed in the section of the table labeled "Unstandardized Coefficients," at the intersection of the column labeled "B" and the row labeled with the predictor variable—in this case, "Educational level (years)." Note that although SPSS uses a capitalized B, the slope is typically symbolized with an italicized lowercase b (this is true for simple and multiple regression). The table shows that the value of the slope of the least-squares regression line is 1,727.53. This value indicates that for every one-year increase in education, we would predict an increase in beginning salary of $1,727.53. The standard error of this slope is displayed next to this value under the column labeled "Std. Error." This value is typically reported along with the value of the slope using the notation SE_b. The standard error of the slope is included in the computation of the t statistic used to determine the statistical significance of the predictor variable.

Coefficients[a]

Model		Unstandardized Coefficients		Standardized Coefficients	t	Sig.
		B	Std. Error	Beta		
1	(Constant)	-6290.967	1340.920		-4.692	.000
	Educational Level (years)	1727.528	97.197	.633	17.773	.000

a. Dependent Variable: Beginning Salary

The value listed in the section of the table labeled "Standardized Coefficients Beta" provides the standardized regression coefficient. It is typically symbolized as β, and since it is a Greek character it should not be italicized when reported. The table displays the value of the standardized slope of the least-squares regression line as .63. The standardized regression coefficient reflects the slope of the least-squares regression line for the standardized (z transformed) variables. Therefore, this value indicates that for every one standard deviation unit increase in educational level, we would predict a .63 standard deviation unit increase in beginning salary. For simple regression, the value of beta will always be the same as the value of the correlation coefficient. You can confirm this for yourself by referring back to the value of the correlation coefficient provided in the "Model Summary" table referred to earlier.

While the F statistic is used to determine the significance of the regression model, the t statistic is used to determine the significance of the predictor variable. You will find the value of the t statistic at the intersection of the column labeled "t" and the row labeled with the name of the predictor variable, in this case the row labeled "Education Level (years)." This value is computed by dividing the unstandardized slope by its

standard error term $\left(t = \dfrac{b}{S_{Eb}} = \dfrac{1,727.5283}{97.1969} = 17.77 \right)$. Consistent with this, the value of t shown in the table

below is 17.77. The adjacent column shows that the p value is less than .001. Once again, since this p value is less than .05, it would typically be considered statistically significant. The degrees of freedom that need to be reported along with the t statistic can be found in the ANOVA table shown previously, in the row labeled "Residual." As shown in that table, the degrees of freedom are equal to 472. Using APA style, we would report these statistics in the following manner: $b = 1,727.53$, $SE_b = 97.20$, $β = .63$, $t(472) = 17.77$, $p < .001$.

Constructing the Equation for the Least-Squares Regression Line

The "Coefficients" table also contains the regression coefficients, which we will need to construct the equation for the least-squares regression line. The equation for the least-squares regression line is $Y' = bX + a$. Alternatively, this formula may be expressed as: $\hat{Y} = a + bX$.

Y' (sometimes denoted $\hat{Y}$) represents the criterion variable. Remember the criterion variable is the variable we want to predict. Again, the unstandardized slope (b) indicates the amount we would predict the criterion (Y) to increase for every one-unit increase in the predictor variable (X). The X in the equation represents the person's score on the predictor variable. We will later substitute in values for X, in order to make our predictions. Finally, the symbol a represents the intercept of the least-squares regression line. It is the place where the least-squares regression line intersects with the Y-axis. It can also be interpreted as the value of Y we would predict for someone with a score of 0 on the predictor variable (X).

As described earlier, the value of the unstandardized slope (b) is displayed at the intersection of the column labeled "B" and the row labeled with the predictor variable ["Educational Level (years)"]. Once again, the table shows the value of the slope is 1,727.53, which indicates that for every one-year increase in education we would predict an increase in beginning salary of $1,727.53. The intercept (a) is presented at the intersection of the column labeled "B" and the row labeled "(Constant)." The table shows that the intercept is $-6{,}290.97$. This value indicates that we would predict a salary of $-\$6{,}290.97$ for a person with 0 years of education (apparently, we would predict that the person is going into debt!) We now have all of the information we need to construct the equation of the least-squares regression line. We simply need to substitute these values of b and a into the equation.

$$Y' = bX + a$$

$$Y' = 1{,}727.53X - 6{,}290.97$$

Once again, the value listed in the column labeled "Standardized Coefficients Beta" provides the standardized regression coefficient, which represents the slope of the least-squares regression line for the standardized (z transformed) variables. You can make predictions of people's Y scores in standardized units (z score units) simply by multiplying their z scores on the X variable by this beta (β) value. That is, $z_{Y'} = \beta z_x$. Importantly, this formula can be used only for standardized variables; only z scores on the X variable can be inputted, and only z scores on the Y variable can be predicted.

Using the Regression Equation to Make Predictions

Now that we have constructed the equation for the least-squares regression line, we can make predictions of people's beginning salaries based on their years of education. What beginning salary would we predict for an individual with 16 years of education? To answer this question all we need to do is substitute 16 (the person's score on the X variable) for X and solve for Y'. Let's go ahead and do that:

$$Y' = 1{,}727.5283(16) - 6{,}290.9673 = 21{,}349.49$$

Thus, we would predict a beginning salary of $\$21{,}349.49$ for a person with 16 years of education. Apparently, these data are quite outdated; let's hope you'll be earning more than that with your four years of postsecondary education! Note that we used values rounded to four decimal places when we were using the regression equation to make a prediction in order to increase the accuracy of our prediction, and we only rounded the final predicted value to two decimal places. While the SPSS output displays only three decimal places, as you learned in Chapter One, the remaining decimal places can be revealed by repeatedly clicking on the value in the table in the output window.

Reporting the Results

There is no single way to report the results of a regression analysis. With that said, the coefficient of determination, the F statistic, the slope (unstandardized and/or standardized), the standard error of the slope, and the t statistic should all be reported. Note that the coefficient of determination should be reported along with the F statistic since it is this value that is evaluated in determining the statistical significance of a model. The beta weight and/or unstandardized slope of the predictor should be reported along with the t statistic. Typically, if the results are being published to address a theoretical question and/or when the unit of measure for the X variable is not well known (e.g., depression scores in the unit of the Beck Depression Inventory), the beta weight is reported. However, if the unit of measure for the X variable is well known (e.g., years of education) and/or the results are being published for the purpose of using the regression model to make predictions, then the unstandardized slope (b values) is reported. We will report both. Using APA style, we could report the following:

> A simple regression analysis was conducted to determine whether beginning salary could be predicted using years of education. Years of education was found to account for 40.09% of the variability in beginning salary, $F(1, 472) = 315.90$, $p < .001$. Accordingly, years of education was a significant predictor of beginning salary, $b = 1{,}727.53$, $SE_b = 97.20$, $\beta = .63$, $t(472) = 17.77$, $p < .001$.

Multiple Regression

Learning Objectives

In this chapter, you will learn how to conduct multiple regression analyses. You will learn how to determine whether regression models, and the individual predictors contained in the models, are statistically significant. You will learn how to interpret the multiple correlation, coefficient of determination, standard error of estimate, as well as partial, semipartial, and squared semipartial correlation coefficients. Finally, you will learn how to use the results of regression analyses to construct the least-squares regression lines and make predictions.

INTRODUCTION TO MULTIPLE REGRESSION

In this chapter, you will learn how to perform multiple regression analyses. Multiple regression is simply an extension of simple regression. Simple regression is used when you want to make predictions using only one predictor variable, while multiple regression is used when you want to make predictions using more than one predictor variable.

We will once again use the sample data set "Employee data" for the demonstrations in this chapter. If you saved the file used for the demonstrations in Chapter Three, you should use it (because we will once again consider the numerically recoded gender variable). If you didn't save the file, the original Employee data file can be found in your SPSS sample files (refer to the "Opening Sample Data Files" section of Chapter Three if you forget how to locate and open these files).

MULTIPLE REGRESSION WITH TWO PREDICTOR VARIABLES

We will begin by considering the simplest case of multiple regression, multiple regression with two predictor variables. Let's say we want to predict people's beginning salaries based on their years of education *and* their previous work experience. Now we are in the land of multiple regression because we are using more than one predictor variable. In this example, beginning salary is the criterion variable (i.e., the variable we want to predict), and it is labeled *Y*, just as it was in simple regression. Educational level and previous experience are the predictor variables (i.e., the variables we will use to make our predictions). The predictor variables are labeled *X*, as in simple regression, but now because there are multiple predictor variables, we need to distinguish them from each other using subscripts. Thus, we will label Educational Level X_1 and Previous Experience X_2.

Conducting a Multiple Regression Analysis (Analyze→Regression→Linear)

To run the analysis, go to **Analyze→Regression→Linear**. A "Linear Regression" dialogue window, like the one shown below, will open. Enter **Beginning Salary** (the criterion variable) into the **Dependent: box** using the corresponding **blue arrow**. Next, enter **Educational Level** and **Previous Experience** (the predictor variables) into the **Independent(s): box** using the corresponding **blue arrow**.

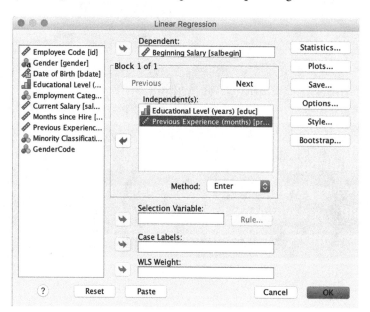

We will also consider the partial and semipartial correlations for the predictor variables since they are commonly used statistics in the context of multiple regression. To obtain these statistics, you need to click on the **Statistics... tab** in the "Linear Regression" dialogue window. This will open the "Linear Regression: Statistics" dialogue window as shown below. Check the **Part and partial correlations** option. Click **Continue** and then click **OK** on the "Linear Regression" window to close the dialogue windows and execute the analysis.

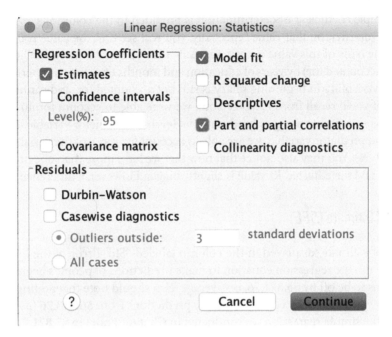

Interpreting the Results
Assessing the Model's Accuracy

We will begin by examining the following 'Model Summary' table.

Model Summary

Model	R	R Square	Adjusted R Square	Std. Error of the Estimate
1	.668[a]	.446	.443	$5,871.763

a. Predictors: (Constant), Previous Experience (months), Educational Level (years)

The Multiple Correlation (*R*)

The multiple correlation is displayed in the column labeled "R." Now that we have multiple predictors, it is appropriate to use a capitalized *R* to symbolize this correlation. The multiple correlation is the correlation between the criterion and the best linear combination of the set of predictor variables. The multiple correlation can also be interpreted as the correlation between the actual *Y* scores (the scores in the data set) and what they would be predicted to be using the regression equation. As you can see in the "Model Summary" table, the value of the multiple correlation is .67. The higher the multiple correlation, the more accurate our predictions will be. This value is quite high, suggesting our predictions will be quite accurate.

The Multiple Coefficient of Determination (R^2)

The multiple coefficient of determination (R^2) is an indicator of the proportion of variability of Y that can be accounted for by the best linear combination of the set of predictors. Once again, in the context of regression it is often interpreted as the proportion of variability of Y that can be predicted by the set of predictors. It is considered an indicator of the effect size.

The value of the multiple coefficient of determination is provided in the column labeled "R Square." As you can see, $R^2 = .45$. If you click on that value repeatedly, you will see that the value rounded to four decimal places is .4458. On the basis of this value we can conclude that 44.58% of the variability in beginning salary can be predicted (or accounted for) by years of education and months of previous experience. That means that about 55.42% of the variability in beginning salary is still not accounted for, and cannot be predicted, using these two variables. If you'll recall from Chapter Four, we were able to account for 40.09% of the variability in beginning salary when we were using only educational level as a predictor variable. By adding in previous experience as a second predictor variable, we are able to account for more of the variability in beginning salary (an additional 4.49%). You may also notice that now that we have more than one predictor, the difference between the R^2 value and the adjusted R^2 value is slightly bigger. However, a difference of .003 is still trivial.

The Standard Error of Estimate (*SEE*)

The standard error of estimate, displayed in the column labeled "Std. Error of the Estimate," is $5,871.76. This means that if we use the regression equation to make predictions of people's beginning salaries, we can expect our predictions to be off by $5,871.76, on average. You should note that adding previous experience as a second predictor variable reduced the error in our predictions from $6,098.26 (as shown in the "Model Summary" table for the simple regression we conducted in Chapter Four) to $5,871.76. In other words, our prediction accuracy has been improved by adding the second predictor variable, previous experience.

Assessing Statistical Significance

The Regression Model

In order to determine whether the regression model is statistically significant (to determine whether we can predict a significant proportion of variability in beginning salary using the set of predictors), we need to consult the following "ANOVA" table. The table clearly shows that the regression model is statistically significant. The value of the F statistic is 189.43, the p value is less than .001, and the degrees of freedom are 2 (see row labeled "Regression") and 471 (see row labeled "Residual"). Therefore, we can report, $R^2 = .45$, $F(2, 471) = 189.43$, $p < .001$.

ANOVA[a]

Model		Sum of Squares	df	Mean Square	F	Sig.
1	Regression	1.306E+10	2	6.531E+9	189.427	.000[b]
	Residual	1.624E+10	471	34477599.6		
	Total	2.930E+10	473			

a. Dependent Variable: Beginning Salary

b. Predictors: (Constant), Previous Experience (months), Educational Level (years)

The Predictors

Next, we need to consider the slopes of the individual predictors and determine whether they are statistically significant. The "Coefficients" table provides this information. The row labeled "Educational Level (years)" provides information about the predictor variable educational level, while the row labeled "Previous Experience (months)" provides information about the predictor variable previous experience.

The unstandardized slopes are displayed in the section of the table labeled "Unstandardized Coefficients," in the column labeled "B." The table shows that the value of the slope for educational level is 1,878.21. This value indicates that for every one-year increase in education, we would predict an increase in beginning salary of $1,878.21, assuming months of previous experience is held constant (i.e., statistically controlled). Notice the value of this slope has changed from the last (simple) regression analysis we conducted using only educational level to predict beginning salary (recall that the value of the slope in the simple regression analysis was 1,727.53). The value of the slope has changed because the influence of previous experience on educational level is controlled for now that previous experience is included in the model. Anytime new predictors are added, the regression coefficients will change.

The "Coefficients" table shows that the value of the unstandardized slope for previous experience is 16.47. This value indicates that for every one-month increase in previous experience, we would predict an increase in beginning salary of $16.47, assuming years of education are held constant. The standard errors of these slopes are displayed next to these values under the column labeled "Std. Error." Once again, these values are typically reported along with the values of the unstandardized slopes using the notation SE_b.

Coefficients[a]

Model		Unstandardized Coefficients		Standardized Coefficients	t	Sig.	Correlations		
		B	Std. Error	Beta			Zero-order	Partial	Part
1	(Constant)	-9902.786	1417.474		-6.986	.000			
	Educational Level (years)	1878.211	96.717	.688	19.420	.000	.633	.667	.666
	Previous Experience (months)	16.470	2.668	.219	6.174	.000	.045	.274	.212

a. Dependent Variable: Beginning Salary

The values listed in the section of the table labeled "Standardized Coefficients Beta" provide the standardized regression coefficients, symbolized as β. The value of the standardized slope for educational level is shown in the table as .69. This value indicates that for every one standard deviation unit increase in educational level, we would predict a .63 standard deviation unit increase in beginning salary, assuming months of previous experience is statistically controlled. The value of the standardized slope for previous experience is .22. This value indicates that for every one standard deviation unit increase in previous experience, we would predict a .22 standard deviation unit increase in beginning salary, assuming years of education are statistically controlled.

Note that unlike simple regression, these values of beta are *not* the same as the values of the correlation coefficients. However, since beta values are standardized, you can directly compare their values in order to determine which variable will be given the most weight when making the predictions. For this reason, these beta values are often referred to as beta weights. Since the beta weight associated with educational level is higher (.69) than the beta weight associated with previous experience (.22), we can determine that people's educational level will be weighted more heavily in making predictions of their beginning salaries than their previous experience. It is not possible to directly compare the unstandardized values of the slopes (the *b* values), since their units of measure will typically differ from one another (indeed in this case, educational level is measured in the unit of years, while previous experience is measured in the unit of months). Comparing unstandardized *b* values is like comparing apples with oranges.

Once again, the *t* statistic is used to determine the significance of the predictor variables. You will find the value of the *t* statistics in the column labeled "t." In this example, educational level has a *t* statistic of

19.42, and previous experience has a t statistic of 6.17. The table also shows that both predictors have p values less than .001, and, therefore, both would be considered statistically significant. This means that each predictor accounts for a significant portion of unique variance in the criterion variable. That is, each predictor accounts for a significant portion of variance in the criterion variable that the other predictor variable does not account for. The degrees of freedom for the t statistics can once again be found in the "ANOVA" table shown previously, in the row labeled "Residual." As shown in that table, the value of the degrees of freedom is 471.

Using all of this information, we can now report that years of education, $b = 1,878.21$, $SE_b = 96.72$, $\beta = .69$, $t(471) = 19.42$, $p < .001$, and months of previous experience, $b = 16.47$, $SE_b = 2.67$, $\beta = .22$, $t(471) = 6.17$, $p < .001$, are statistically significant predictors of beginning salary.

Coefficients[a]

Model		Unstandardized Coefficients B	Std. Error	Standardized Coefficients Beta	t	Sig.	Correlations Zero-order	Partial	Part
1	(Constant)	-9902.786	1417.474		-6.986	.000			
	Educational Level (years)	1878.211	96.717	.688	19.420	.000	.633	.667	.666
	Previous Experience (months)	16.470	2.668	.219	6.174	.000	.045	.274	.212

a. Dependent Variable: Beginning Salary

Constructing the Equation for the Least-Squares Regression Line

We can also use the information in the "Coefficients" table shown above to construct the equation for the least-squares regression line. The multiple regression equation for two predictor variables is: $Y' = b_1X_1 + b_2X_2 + a$. Alternatively, this formula may be expressed as: $\hat{Y} = a + bX_1 + bX_2$. The equations are very similar to the simple regression equations; they have simply been expanded to include two predictor variables (X_1 and X_2) and each of their slopes (b_1 and b_2).

The values of the intercept (a) and slopes (b_1 and b_2) are presented in the column labeled "B." As described in the previous section, the table shows that b_1 is 1,878.21 and b_2 is 16.47. The intercept is presented at the intersection of the column labeled "B" and the row labeled "(Constant)." The value of the intercept is $-9,902.79$. This value indicates that we would predict a beginning salary of $\$-9,902.79$ for a person with 0 years of education and 0 months of previous experience. We now have all of the information we need to construct the equation of the least-squares regression line. We simply need to substitute these values in for b_1, b_2, and a.

$$Y' = b_1X_1 + b_2X_2 + a = 1,878.21X_1 + 16.47X_2 - 9,902.79$$

Using the Regression Equation to Make Predictions

Since both of our predictor variables are statistically significant, we can make predictions of people's beginning salaries based on their years of education and months of previous experience. What beginning salary would we predict for an individual with 16 years of education and 110 months of previous experience? To answer this question, all we need to do is substitute 16 for X_1 and 110 for X_2 and then solve for Y'. Let's go ahead and do that:

$$Y' = 1,878.2114(16) + 16.4704(110) - 9,902.7861 = 21,960.34$$

Thus, we would predict a beginning salary of $21,960.34 for a person with 16 years of education and 110 months of previous experience. Once again, please note that we used values rounded to four decimal places when we were using the regression equation to make a prediction in order to increase the accuracy of our prediction, and we rounded our final answer to only two decimal places.

Bivariate, Partial, Semipartial, and Squared Semipartial Correlations

Since we selected the option to compute the part and partial correlations before executing the analysis, the "Coefficients" table will also include a section labeled "Correlations." This section includes the values of the bivariate, partial, and semipartial correlations between each of the predictors and the criterion.

Coefficients[a]

Model		Unstandardized Coefficients		Standardized Coefficients	t	Sig.	Correlations		
		B	Std. Error	Beta			Zero-order	Partial	Part
1	(Constant)	-9902.786	1417.474		-6.986	.000			
	Educational Level (years)	1878.211	96.717	.688	19.420	.000	.633	.667	.666
	Previous Experience (months)	16.470	2.668	.219	6.174	.000	.045	.274	.212

a. Dependent Variable: Beginning Salary

Bivariate Correlations (*r*)

The column labeled "Zero-order" displays the bivariate correlations, which are simply the correlations between each of the predictors and the criterion. As established previously, the correlation between educational level and beginning salary is equal to .63. The table also shows that the correlation between previous experience and beginning salary is equal to .05.

Partial Correlations (*pr*)

If you'll recall from Chapter Three, a partial correlation is a correlation between two variables, after statistically controlling for a potential third variable. The column labeled "Partial" provides the partial correlations between each of the predictors and the criterion after statistically controlling for the other predictor variable. More precisely, the partial correlation for X_1 can be interpreted as the relationship between X_1 and Y after the influence of X_2 on both X_1 and Y has been removed. So it is the relationship between the predictor and the criterion after the influence of the other predictor on both the predictor of interest and the criterion has been removed.

As you can see, the partial correlation for educational level is .67. This means that the correlation between educational level and beginning salary, after the influence of previous experience on both educational level and beginning salary has been removed, is. 67. Since there is little difference between the bivariate and partial correlations, we can determine that previous experience has little to no influence on the relationship between educational level and beginning salary. (If this is not clear, then review the section on "Interpreting Partial Correlations" in Chapter Three.)

The partial correlation for previous experience is .27. This means that the correlation between previous experience and beginning salary, after the influence of educational level on both previous experience and beginning salary has been removed, is .27. You may be surprised to see that controlling for the influence of years of education increased the size of the correlation between previous experience and beginning salary. When a partial (or semipartial) correlation is higher than the original bivariate correlation, it indicates the presence of a suppression effect. Suppression is an advanced statistical concept, and suppression effects can be quite difficult to interpret. Basically, partial correlations involve removing variability in the predictor and/or criterion that is accounted for (or predicted) by the other predictor(s). When we see a suppression effect like this, it suggests that by controlling for the other predictor variable(s), we are removing some error variance (variability in Y and/or X unrelated to each other) and in doing so we are making our variable a better predictor.

Semipartial Correlations (*sr*)

SPSS refers to semipartial correlations by their old-school name, part correlation. The semipartial correlation is a close cousin of the partial correlation. It also provides an indicator of the relationship between two variables, after statistically controlling for a potential third variable. However, the degree of control differs somewhat. For partial correlation, the influence of the potential third variable on *both* the predictor and the criterion variable is statistically controlled. In contrast, for semipartial correlation, the influence of the potential third variable on *only* the other predictor variable is controlled. Precisely, the semipartial correlation for X_1 can be interpreted as the relationship between X_1 and Y after the influence of X_2 on X_1 has been removed. The notation for the semipartial correlation is sr.

The semipartial correlation (.67) for educational level displayed in the table indicates that the correlation between educational level and beginning salary, after the influence of previous experience on educational level has been removed, is .67. The fact that the bivariate and semipartial correlations for educational level do not differ much from one another suggests there is little relationship between previous experience and educational level (since controlling for the influence of previous experience on educational level does little to the magnitude of the correlation). Finally, the semipartial correlation for previous experience indicates that the correlation between previous experience and beginning salary, after the influence of educational level on previous experience has been removed is .21. Once again, we see a suppression effect at play here.

Squared Semipartial Correlations (*sr²*)

The squared semipartial correlation, sr^2, is a very informative statistic in the context of multiple regression. Its value can be interpreted as the proportion of variability in the criterion that is uniquely associated with the predictor, or that can uniquely be predicted by the predictor. In other words, it is an indicator of the proportion of the criterion (the variable we are trying to predict) that only that specific predictor variable can account for (or predict). As such, this statistic provides an indicator of the unique contribution of a predictor variable to a regression model. More specifically, the squared semipartial correlation tells us how much R^2 will decrease if that predictor variable is removed from the regression model (or how much R^2 has increased as a result of including the predictor in the model).

The squared semipartial correlation is simply the semipartial correlation squared. Thus, the squared semipartial correlation for the predictor variable educational level is .4437 ($.6661^2 = .4437$). This value indicates that educational level uniquely predicts 44.37% of the variability of beginning salary and that if the variable educational level was removed from the regression model, the value of R^2 would decrease by .44. The squared semipartial correlation for the predictor variable previous experience is .0449 ($.2118^2 = .0449$). This value indicates that previous experience uniquely predicts about 4.49% of the variability of beginning salary and that if previous experience was removed from the regression model, the value of R^2 would decrease by only .04. While this decrease seems trivial, the fact that previous experience is a significant predictor of beginning salary suggests that this decrease would be statistically significant.

Reporting the Results

We could report these results in the following manner:

> A multiple regression analysis was used to predict beginning salary using years of education and months of previous experience. These two variables were found to account for a significant

proportion of variability in beginning salary, $R^2 = .45$, $F(2, 471) = 189.43$, $p < .001$. Moreover, both years of education, $b = 1,878.21$, $SE_b = 96.72$, $\beta = .69$, $t(471) = 19.42$, $p < .001$, and months of previous experience, $b = 16.47$, $SE_b = 2.67$, $\beta = .22$, $t(471) = 6.17$, $p < .001$, were found to be significant predictors of beginning salary.

MULTIPLE REGRESSION WITH A NOMINAL PREDICTOR WITH TWO CATEGORIES

To illustrate multiple regression with three predictor variables, one of which is a nominal variable with two categories (a dichotomous predictor variable), we will predict beginning salary using educational level, previous experience, and gender. We will label Educational Level X_1, Previous Experience X_2, and Gender X_3.

Gender is a nominal predictor variable with two categories (men, women). Since it was originally entered as a string (text) variable, we will not be able to perform any analyses with it until it has been recoded using a numeric code. If you are using the data set you used in Chapter Three, then gender has already been numerically coded as "GenderCode." If you are not using the data set you used in Chapter Three with the numerically recoded gender variable, then you will need to recode the gender variable. To recode the gender variable, go to **Transform→Recode into Different Variables**. Name the recoded variable **Gender-Code,** and then proceed to recode the current gender code "**m**" to "**1**" and "**f**" to "**2**" (refer to the section entitled "Recoding Variables" in Chapter One, if you forget how to do this).

Conducting a Multiple Regression Analysis (Analyze→Regression→Linear)

To conduct the multiple regression analysis, you will need to go to **Analyze→Regression→Linear.** This will open the following "Linear Regression" dialogue window. Beginning Salary (the criterion variable) should still appear in the Dependent: box, and Educational Level and Previous Experience should still appear in the Independent(s): box, from the last analysis we conducted. If they do not, then you will need to move them into these boxes. Finally, enter **GenderCode** into the **Independent(s): box** using the corresponding **blue arrow**. All three predictor variables should now appear in the Independent(s): box.

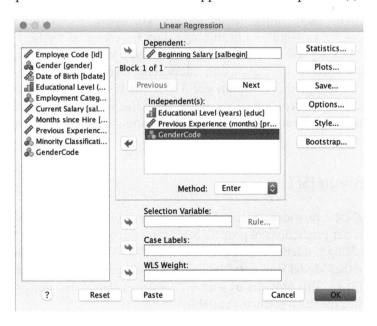

Once again, click on the **Statistics… tab** to open the "Linear Regression: Statistics" dialogue widow. Make sure that the option for **Part and partial correlations** is still checked. Click **Continue** and then **OK** to close both dialogue windows and execute the analysis.

Interpreting the Results
Assessing the Model's Accuracy

We will begin with the "Model Summary" table shown here.

Model Summary

Model	R	R Square	Adjusted R Square	Std. Error of the Estimate
1	.696[a]	.484	.481	$5,671.155

a. Predictors: (Constant), GenderCode, Previous Experience (months), Educational Level (years)

The Multiple Correlation (R)

The column labeled "R" shows the multiple correlation is .70. Therefore, the correlation between beginning salary and the best linear combination of educational level, previous experience, and gender is .70. Alternatively, this value can be interpreted as the correlation between actual Y scores and what we would predict them to be using the regression equation. The value of the multiple correlation has increased slightly by adding the third predictor variable, suggesting that the accuracy of our prediction has been slightly improved.

The Multiple Coefficient of Determination (R^2)

The "R Square" value in the table shows that the multiple coefficient of determination is .48. This value indicates that 48.41% of the variability in beginning salary is accounted for (and can be predicted by) years of education, months of previous experience, and gender. If you'll recall, we were able to account for 44.58% of the variability when we were using only educational level and previous experience as predictor variables. This means we can predict an additional 3.83% (48.41–44.58) of the variability in beginning salary by including gender as a predictor. (If you understood the previous section on squared semipartial correlations you should already be able to determine that the squared semipartial correlation for the coded gender variable will be .0383.)

The Standard Error of Estimate (*SEE*)

The "Model Summary" table provides a value of $5,671.16 for the standard error of estimate. This means that we can expect that our predictions of people's beginning salaries will be off by $5,671.16, on average. You should note that adding gender as a third predictor variable reduced the error in our predictions from $5,871.76 (as shown in the "Model Summary" table in the "Multiple Regression with Two Predictor Variables" section) to $5,671.16. This provides us with even more evidence that our prediction accuracy has been improved by adding the third predictor variable.

Assessing Statistical Significance
The Regression Model

In order to determine whether the regression model is statistically significant (to determine whether we can reliably predict beginning salary using the set of three predictors) we need to consult the "ANOVA" table shown below. The table clearly shows that the regression model is statistically significant. As shown in the table, the value of the F statistic is 147.01, the p value is less than .001, and the degrees of freedom are 3 and 470. Once again, this would be reported as: $R^2 = .48$, $F(3, 470) = 147.01$, $p < .001$.

ANOVA[a]

Model		Sum of Squares	df	Mean Square	F	Sig.
1	Regression	1.418E+10	3	4.728E+9	147.014	.000[b]
	Residual	1.512E+10	470	32162000.3		
	Total	2.930E+10	473			

a. Dependent Variable: Beginning Salary

b. Predictors: (Constant), GenderCode, Previous Experience (months), Educational Level (years)

The Predictors

The unstandardized slopes are displayed in the section of the table labeled "Unstandardized Coefficients," in the column labeled "B." The table displays the value of the slope for educational level as 1,625.29. This value indicates that for every one-year increase in education, we would predict an increase in beginning salary of $1,625.29, assuming previous experience and gender are statistically controlled. The table displays the value of the slope for previous experience as 12.00. This value indicates that for every one-month increase in previous experience, we would predict an increase in beginning salary of $12.00, assuming years of education and gender are held constant. The value of the slope for gender is –3,446.50. The negative value indicates an inverse slope. In other words, the negative slope for gender means that the relationship between the predictor and the criterion is negative (i.e., that as the value of the predictor variable increases, the value of the criterion variable decreases). Since gender is a nominal variable and we used numeric codes to represent the values of this variable, we need to carefully consider the codes that we used, before we can understand the nature of this relationship. If you'll recall, we used a code of 1 to represent men and a code of 2 to represent women. As such, the negative slope associated with this variable indicates that being a woman (a higher value) is associated with lower beginning salaries and, similarly, that being a man (a lower value) is associated with higher beginning salaries. By considering the value of the slope, we can determine that being a woman is associated with a decrease in beginning salary of $3,446.50 (assuming educational level and previous experience are held constant).

Coefficients[a]

Model		Unstandardized Coefficients B	Std. Error	Standardized Coefficients Beta	t	Sig.	Correlations Zero-order	Partial	Part
1	(Constant)	-1045.042	2030.198		-.515	.607			
	Educational Level (years)	1625.292	102.753	.596	15.817	.000	.633	.589	.524
	Previous Experience (months)	12.001	2.685	.159	4.469	.000	.045	.202	.148
	GenderCode	-3446.504	583.307	-.218	-5.909	.000	-.457	-.263	-.196

a. Dependent Variable: Beginning Salary

The values listed in the section of the table labeled "Standardized Coefficients Beta" provide the standardized regression coefficients, symbolized as β. The table displays the value of the standardized slope for educational level as .60. This value indicates that for every one standard deviation unit increase in educational level, we would predict a .60 standard deviation unit increase in beginning salary, assuming previous experience and gender are statistically controlled. The value of the standardized slope for previous experience is .16. This value indicates that for every one standard deviation unit increase in previous experience, we would predict a .16 standard deviation unit increase in beginning salary, assuming years of education and gender are statistically controlled. The standardized slope for gender is −.22, indicating that for every one standard deviation unit increase in gender, we would predict a .22 standard deviation unit decrease in beginning salary, assuming years of education and months of previous experience are statistically controlled. Since a one standard deviation unit increase in gender is difficult to interpret, the slope for this nominal variable has a less meaningful interpretation. However, by comparing the size of these three beta weights, we can determine that educational level is still the strongest predictor. (It has the highest beta weight and will therefore be weighted most in making our predictions.) Gender has the second highest beta weight and will therefore be weighted less than educational level but more than previous experience in making our predictions. Finally, previous experience has the lowest beta weight and will therefore be given the least weight in our predictions.

Once again, the t statistic is used to determine the significance of the predictor variables. You will find the value of the t statistics in the column labeled "t." In this example, educational level has a t statistic of 15.82, previous experience has a t statistic of 4.47, and gender has a t statistic of −5.91. The table also shows that all three predictors have p values less than .001, and therefore all three would be considered statistically significant. This means that each predictor accounts for a significant portion of unique variance in the criterion variable. That is, each predictor accounts for a significant portion of variance in the criterion variable that the other predictor variable does not account for. The degrees of freedom for the t statistics can once again be found in the "ANOVA" table shown previously, in the row labeled "Residual." As shown in that table, the value of the degrees of freedom is 470.

Using all this information, we can now report that years of education, $b = 1625.29$, $SE_b = 102.75$, $β = .60$, $t(470) = 15.82$, $p < .001$; months of previous experience, $b = 12.00$, $SE_b = 2.69$, $β = .16$, $t(470) = 4.47$, $p < .001$; and gender, $b = −3446.50$, $SE_b = 583.31$, $β = −.22$, $t(470) = −5.91$, $p < .001$; are all statistically significant predictors of beginning salary.

Coefficients[a]

Model		Unstandardized Coefficients B	Std. Error	Standardized Coefficients Beta	t	Sig.	Correlations Zero-order	Partial	Part
1	(Constant)	−1045.042	2030.198		−.515	.607			
	Educational Level (years)	1625.292	102.753	.596	15.817	.000	.633	.589	.524
	Previous Experience (months)	12.001	2.685	.159	4.469	.000	.045	.202	.148
	GenderCode	−3446.504	583.307	−.218	−5.909	.000	−.457	−.263	−.196

a. Dependent Variable: Beginning Salary

Constructing the Equation for the Least-Squares Regression Line

The preceding "Coefficients" table also contains all of the values we need to construct the equation for the least-squares regression line. The equation used for a model with three predictor variables is $Y' = b_1X_1 + b_2X_2 + b_3X_3 + a$. Alternatively, this formula may be expressed as $\hat{Y} = a + bX_1 + bX_2 + bX_3$. This equation is very similar to the equation for multiple regression with two predictor variables; it has simply been expanded to include a third predictor variable (X_3) and its slope (b_3).

Let's begin by finding the values of the regression coefficients (slopes and intercept). Once again, they are all presented in the column labeled "B." The table shows that b_1 (the slope for educational level) is 1,625.29, b_2 (the slope for previous experience) is 12.00, and b_3 (the slope for gender) is $-3,446.50$. The intercept, shown at the intersection of the column labeled "B" and the row labeled "(Constant)," is $-1,045.04$. We now have all of the information we need to construct the equation of the least-squares regression line. All we need to do is substitute in the values of b_1, b_2, b_3, and a.

$$Y' = 1,625.29X_1 + 12.00X_2 - 3,446.50X_3 - 1,045.04$$

Using the Regression Equation to Make Predictions

Since all of our predictors are statistically significant, we can now use the regression equation to predict people's beginning salaries based on their years of education, months of previous experience, and gender[1]. Let's say that Joe (a man) has 16 years of education and 80 months of previous experience. What would we predict his beginning salary to be? To answer this question, we simply need to substitute these values into the corresponding X values in our regression equation and then solve the equation. Since we labeled educational level X_1, previous experience X_2, and gender X_3, we need to substitute 16 for X_1, 80 for X_2, and 1 (the numeric code we have assigned for men) for X_3. Let's go ahead and do that:

$$1,625.2918(16) + 12.0014(80) - 3,446.5036(1) - 1,045.0423 = 22,473.23$$

Thus, we would predict Joe's beginning salary to be $22,473.23. Try to predict what Jane's beginning salary would be if she had the same number of years of education and the same months of previous experience as Joe. You will discover that her beginning salary would be predicted to be $3,446.50 less than Joe's salary (which is the value of b_3).

Bivariate, Partial, Semipartial, and Squared Semipartial Correlations

The "Correlations" section of the "Coefficients" table also shows the bivariate, partial, and semipartial correlations for all three predictor variables.

Coefficients[a]

Model		Unstandardized Coefficients B	Std. Error	Standardized Coefficients Beta	t	Sig.	Correlations Zero-order	Partial	Part
1	(Constant)	−1045.042	2030.198		−.515	.607			
	Educational Level (years)	1625.292	102.753	.596	15.817	.000	.633	.589	.524
	Previous Experience (months)	12.001	2.685	.159	4.469	.000	.045	.202	.148
	GenderCode	−3446.504	583.307	−.218	−5.909	.000	−.457	−.263	−.196

a. Dependent Variable: Beginning Salary

Bivariate Correlations (*r*)

The column labeled "Zero-order" shows that the bivariate correlation between educational level and beginning salary is .63, the correlation between previous experience and beginning salary is .05, and the correlation between gender and beginning salary is –.46. Once again, since we coded men with 1s (a lower

---

[1] If one of the predictor variables were not statistically significant, then we would need to rerun the regression analysis without including it as a predictor variable before constructing the equation for the regression line and making predictions.

value) and women with 2s (a higher value), this negative correlation indicates that being a woman is associated with earning a lower beginning salary.

Partial Correlations (*pr*)

The column labeled "Partial" shows the partial correlations for all three predictor variables. In the case of a model with three predictors, the partial correlation for X_1 can be interpreted as the relationship between X_1 and Y, after the influences of X_2 and X_3 on both X_1 and Y have been removed. So it is the relationship between the predictor and the criterion after the influences of the other predictor variables on both the predictor of interest and the criterion have been removed. The partial correlation of .59 for educational level provided in the table indicates that the correlation between educational level and beginning salary, after the influences of previous experience and gender on both educational level and beginning salary have been removed, is. 59. The partial correlation of .20 displayed for previous experience indicates that the correlation between previous experience and beginning salary, after the influences of educational level and gender on both previous experience and beginning salary have been removed, is. 20. Finally, the partial correlation of –.26 for the gender variable indicates that the correlation between gender and beginning salary, after the influences of educational level and previous experience on both gender and beginning salary have been removed, is. –.26.

Semipartial Correlations (*sr*)

The column labeled "Part" shows the semipartial correlations for all three predictor variables. In the case of a regression model with three predictors, the semipartial correlation for X_1 can be interpreted as the relationship between X_1 and Y after the influences of X_2 and X_3 on X_1 have been removed. So it is the relationship between the predictor and the criterion after the influence of the other predictors on just the predictor of interest have been removed.

The semipartial correlation of .52 for educational level displayed in the "Part" column of the "Coefficients" table indicates that the correlation between educational level and beginning salary, after the influences of previous experience and gender on educational level have been removed, is. 52. The semipartial correlation of .15 displayed for previous experience indicates that the correlation between previous experience and beginning salary, after the influence of educational level and gender on previous experience have been removed, is .15. Finally, the semipartial correlation of –.20 provided for the gender variable informs us that the correlation between gender and beginning salary, after the influence of educational level and previous experience on gender have been removed, is –.20.

Squared Semipartial Correlations (*sr*^{2})

To review, the squared semipartial correlation indicates the proportion of variability in the criterion that the predictor (and only that predictor) can predict. In other words, it is an indicator of the unique contribution of a predictor variable to a regression model, and therefore its value tells us how much R^2 will decrease if that predictor variable is removed from the regression model. Similarly, it informs us how much R^2 has increased as a result of the inclusion of the predictor.

The squared semipartial correlation for educational level is .27 ($.5240^2 = .2746 = .27$). This value indicates that educational level uniquely predicts 27.46% of the variability of beginning salary (i.e., 27.46% of the variability in beginning salary is being predicted by educational level alone). Therefore, if the variable educational level was removed from the regression model, the value of R^2 would decrease by .27. The squared semipartial correlation for previous experience is .02 ($.1481^2 = .0219 = .02$). This value indicates that previous experience uniquely predicts 2.19% of the variability of beginning salary, and that if the variable previous experience was removed from the regression model, the value of R^2 would decrease by only .02. Finally, the squared semipartial correlation for the gender variable is .04 ($-.1958^2 = .0383 = .04$). This value indicates that of all of the variability in beginning salary, 3.83% of it is being predicted uniquely by gender.[2]

Reporting the Results

An example of an APA style write-up of these results follows:

A multiple regression analysis was used to predict beginning salary using years of education, months of previous experience, and gender. The three predictors were found to account for just under half of the variability in beginning salary ($R^2 = .48$), which was statistically significant, $F(3, 470) = 147.01$, $p < .001$. Years of education, $b = 1,625.29$, $SE_b = 102.75$, $\beta = .60$, $t(470) = 15.82$, $p < .001$; months of previous experience, $b = 12.00$, $SE_b = 2.69$, $\beta = .16$, $t(470) = 4.47$, $p < .001$; and gender, $b = -3,446.50$, $SE_b = 583.31$, $\beta = -.22$, $t(470) = -5.91$, $p < .001$; were all found to be significant predictors of beginning salary, with the negative coefficient for gender indicating that women earn significantly lower beginning salaries than men. An examination the squared semipartial correlations revealed that education uniquely predicts 27.46% of the variability in beginning salary, 3.83% of the variance in beginning salary can be uniquely predicted by gender, and previous experience uniquely accounts for only 2.19% of the variance in beginning salaries.

MULTIPLE REGRESSION WITH A NOMINAL PREDICTOR WITH MORE THAN TWO CATEGORIES

To illustrate multiple regression with a nominal variable with more than two categories, we will try to predict beginning salary using educational level, previous experience, gender, and job category. Job category is a nominal variable with three categories (clerical, custodial, and managerial). Once again, only dichotomous nominal variables (i.e., nominal variables with two categories) can be meaningfully analyzed using correlation or regression, so before we can use this variable in a regression analysis, we first need to break it into a series of dichotomous variables. More specifically, we will need to break the nominal variable into $g - 1$ variables, where g refers to the number of categories. Since there are three categories in the job category variable, we will need to create two new variables ($3 - 1 = 2$). Specifically, we will create a Custodial variable and a Managerial variable, and we will use Clerical (the most frequent category) as our reference group. If there were a fourth category (Administrative), then we would need to create three new variables ($4 - 1 = 3$), and so on.

Creating New Dichotomous Variables

Let's start by creating a new variable named "Custodial." First, we will need to see how the original variable was coded. Go to **Variable View** and click on **Values** in the row labeled **jobcat**. As illustrated below, you will see that Clerical staff are coded with a 1, Custodians are coded with a 2, and Managerial staff are coded with a 3. You might also notice that missing data are coded with a 0, but we don't need to worry about that

---

[2] You may notice that the sum of all of the squared semipartial correlations does not equal the value of R^2. This is because some of the value of R^2 is made up of shared, rather than unique, variance (it reflects variability of Y that can be predicted by more than one predictor variable).

since that missing data code is also indicated in the Missing column in Variable View (as such, any 0s will simply be ignored and treated as missing data).

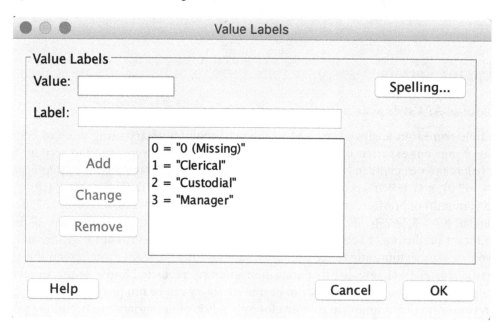

Next, go to **Transform → Recode into Different Variables.** Move the Employment Category (jobcat) variable in the **Numeric Variable – > Output Variable: box** using the **blue arrow**. Type the name of the new variable you want to create, in this case **Custodial**, in the **Output Variable Name: box,** and then click **Change**. Next, click the **Old and New Values… tab.**

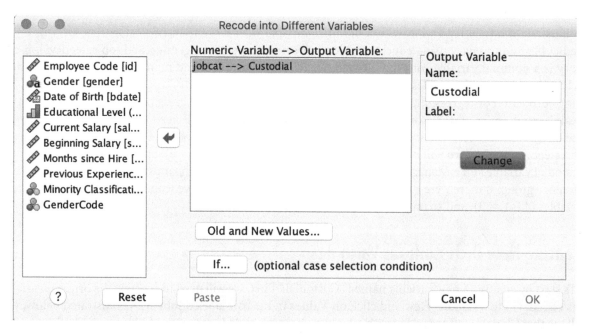

Clicking this tab will open a new "Recode into Different Variables: Old and New Values" dialogue window, which we will use to code the levels of this new dichotomous variable. Specifically, we will recode Custodians with 1s, and we will recode Clerical staff and Managers with 0s (to indicate that they are not Custodians). Since Custodians were originally coded with 2s, we will need to recode 2s as 1s. To do this we simply need to enter a **2** in the **Old Value Value: box** and a **1** in the **New Value Value: box and** then click **Add.**

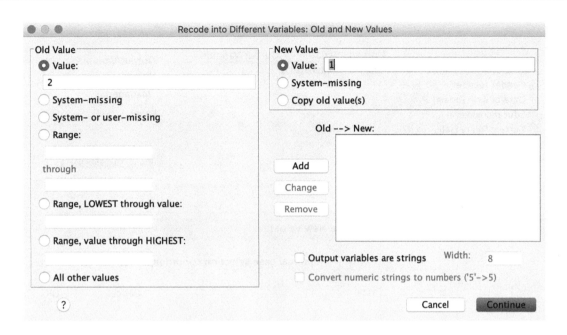

Next, we need to recode Clerical staff with 0s, to indicate that they are not Custodians. To do this, simply enter 1 (the original code for Clerical staff) into the **Old Value Value: box** and **0** in the **New Value Value: box.** Then click **Add**. Finally, to recode the Managerial staff with 0s, to indicate they are not Custodians, enter 3 (the original code for Managerial staff) in the **Old Value Value: box** and **0** in the **New Value Value: box.** Then click **Add**. Your dialogue window should now appear like the one shown below. Finally, click **Continue** and then click **OK** on the "Recode into Different Variables" dialogue window, to create the new dichotomous Custodial variable.

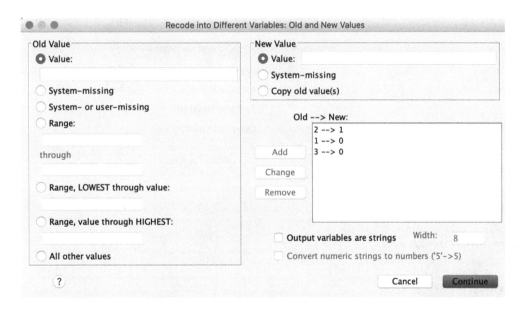

We will now repeat this process to create a Managerial variable, for which Managerial staff will be coded with 1s and Custodial and Clerical staff will be coded with 0s (to indicate they are not Managers). Once again, go to **Transform → Recode into Different Variables**. The employment category (jobcat) variable should still appear in the **Numeric Variable – > Output Variable: box**. Delete the label Custodial in the **Output Variable Name: box,** and replace it with the label **Managerial.** Then click **Change.** Next, click the option for **Old and New Values…**

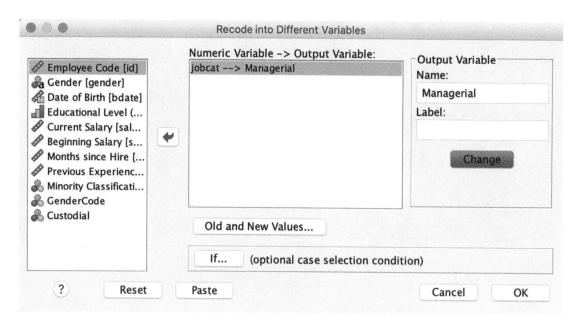

As shown below, clicking this option will open a new dialogue window that we will use to code the levels of this new dichotomous variable. First, we will need to remove the last coding scheme we used by clicking on each item in the **Old -- > New: box** and clicking **Remove** until the box is empty. Next, since Managerial staff were originally coded with 3s, we will need to recode 3s as 1s. To do this, we simply need to enter a **3** in the **Old Value Value: box** and a **1** in the **New Value Value: box**, then click **Add**. Custodians were originally coded with 2s, so we will need to recode 2s as 0s. Enter a **2** in the **Old Value Value: box** and a **0** in the **New Value Value: box**, then click **Add**. Finally, Clerical staff were originally coded with 1s, so we will need to recode 1s as 0s. Enter a **1** in the **Old Value Value: box** and a **0** in the **New Value Value: box**, then click **Add**. Finally click **Continue** to create the new dichotomous "Managerial" variable.

Conducting a Multiple Regression Analysis (Analyze→Regression→Linear)

To conduct the analysis, you will need to go to **Analyze→Regression→Linear.** This will open the "Linear Regression" dialogue window shown below. Beginning Salary (the criterion variable) should still appear in the Dependent: box, and Educational Level, Previous Experience, and GenderCode should still appear in the Independent(s): box, from the last analysis we conducted. If they do not, then you will need to move them into these boxes. Next, enter the new dichotomous variables we created – **Custodial** and **Managerial** – into the **Independent(s): box** using the **blue arrow**. Your "Linear Regression" dialogue window should now look like the one displayed here.

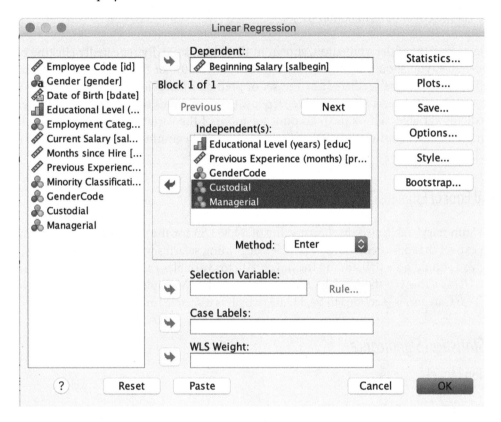

Once again, click on the **Statistics tab** to open the "Linear Regression: Statistics" dialogue widow. Make sure the **Part and partial correlations** option is checked. Click **Continue** and then **OK** on the "Linear Regression" window to close both dialogue windows and execute the analysis.

Interpreting the Results

Assessing the Model's Accuracy

The Multiple Correlation (R)

By referring to the column labeled "R" in the "Model Summary" table shown here, you should be able to see that the multiple correlation is .84. There has been a reasonable increase in this value since adding the Custodial and Managerial variables as predictor variables. This suggests that the accuracy of our prediction has been improved from that obtained in the previous model that we considered.

Model Summary

Model	R	R Square	Adjusted R Square	Std. Error of the Estimate
1	.837[a]	.700	.697	$4,333.832

a. Predictors: (Constant), Managerial, Previous Experience (months), GenderCode, Custodial, Educational Level (years)

The Multiple Coefficient of Determination (R^2)

The multiple coefficient of determination, shown in the table, is .70. By repeatedly clicking on this value in the output table, you should be able to determine that 70.00% of the variability in beginning salary is accounted for (and can be predicted) by the set of predictor variables. If you'll recall, we were able to account for 48.81% of the variability when we were using the previous set of predictors. Therefore, adding the new predictor variables allows us to account for more of the variability in beginning salary. Specifically, we can predict an additional 21.19% of the variability in beginning salary by including job category in the model.

The Standard Error of Estimate (*SEE*)

The "Model Summary" table also displays a value of $4,333.83 for the standard error of estimate. Therefore, we can expect that our predictions of people's beginning salaries will be off by $4,333.83, on average. Once again, you should note that adding the new predictor variables reduced the error in our predictions from $5,671.16 (as shown in the Model Summary table in the "Multiple Regression with a Nominal Predictor with Two Categories" section) to $4,333.83, which is a sizable improvement.

Assessing Statistical Significance

The Regression Model

Next we will consider the "ANOVA" table shown here to determine whether we can reliably predict people's beginning salaries using the set of predictors. The table clearly shows that the regression model is statistically significant. The value of the F statistic is 218.41, the p value is less than .001 and the degrees of freedom that we will need to report are 5 and 468. Once again, this would be reported as: $R^2 = .70$, $F(5, 468) = 218.41$, $p < .001$.

ANOVA[a]

Model		Sum of Squares	df	Mean Square	F	Sig.
1	Regression	2.051E+10	5	4.102E+9	218.409	.000[b]
	Residual	8.790E+9	468	18782100.5		
	Total	2.930E+10	473			

a. Dependent Variable: Beginning Salary

b. Predictors: (Constant), Managerial, Previous Experience (months), GenderCode, Custodial, Educational Level (years)

The Predictors

Next, we need to consider the individual predictors. First, the "Coefficients" table, shown below, indicates that the value of the unstandardized slope for educational level is 664.92. This value indicates that for every one-year increase in education, we would predict an increase in beginning salary of $664.92, assuming previous experience, gender, and job category are statistically controlled. The table displays the value of the unstandardized slope for previous experience as 10.58. This value indicates that for every one-month increase in previous experience, we would predict an increase in beginning salary of $10.58, assuming years of education, gender, and job category are held constant. The value of the slope for gender is –2,659.89. Again, the negative value indicates an inverse slope, indicating that being a woman (a higher value) is associated with lower beginning salaries and, similarly, that being a man (a lower value) is associated with higher beginning salaries. By considering the value of the slope, we can determine that being a woman is associated with a decrease in beginning salary of $2,659.89 (assuming educational level, previous experience, and job category are controlled for). The value of the slope for our new custodial variable is –998.90, which indicates that custodial staff earn, on average, $998.90 less than clerical staff (our reference group), even after statistically controlling for differences in years of education, months of previous experience, and gender. Finally, the value of the slope for the new managerial variable is 12,133.70, which indicates that managerial staff earn, on average, $12,133.70 more than clerical staff, even after statistically controlling for differences in years of education, months of previous experience, and gender.

Coefficients[a]

Model		Unstandardized Coefficients		Standardized Coefficients	t	Sig.	Correlations		
		B	Std. Error	Beta			Zero-order	Partial	Part
1	(Constant)	8809.497	1706.486		5.162	.000			
	Educational Level (years)	664.916	96.781	.244	6.870	.000	.633	.303	.174
	Previous Experience (months)	10.583	2.215	.141	4.778	.000	.045	.216	.121
	GenderCode	-2659.888	466.598	-.168	-5.701	.000	-.457	-.255	-.144
	Custodial	-998.898	1035.451	-.029	-.965	.335	-.061	-.045	-.024
	Managerial	12133.699	661.834	.589	18.333	.000	.782	.647	.464

a. Dependent Variable: Beginning Salary

The "Coefficients" table also displays the value of the standardized slopes in the column labeled "Standardized Coefficients Beta." Specifically, the table shows the value of beta for educational level as .24. This value indicates that for every one standard deviation unit increase in educational level, we would predict a .24 standard deviation unit increase in beginning salary, assuming previous experience, gender, and job category are all statistically controlled. The value of the standardized slope for previous experience is .14. This value indicates that for every one standard deviation unit increase in previous experience, we would predict a .14 standard deviation unit increase in beginning salary, assuming years of education, gender, and job category are statistically controlled. Since a one standard deviation unit increase in gender, custodial, and managerial is difficult to interpret, the slopes for these nominal variables have less meaningful interpretations. However, by comparing the size of these beta weights, we can determine that managerial is now the strongest predictor. (It has the highest beta weight and will therefore be weighted most in making our predictions.) Educational level has the second highest beta weight, gender has the third highest beta weight, previous experience now has the fourth highest beta weight, and custodial has the lowest beta weight and will therefore be given the least weight in our predictions.

Once again, the t statistics are used to determine whether the predictors are statistically significant. In this example, educational level has a t statistic of 6.87, previous experience has a t statistic of 4.78, gender has a t statistic of –5.70, custodial has a t statistic of –0.96, and managerial has a t statistic of 18.33. The table also shows that the three predictors we considered previously have p values less than .001. As indicated previously, this means that education, previous experience, and gender are statistically significant predictors of beginning salary (each predicts a significant portion of unique variance in beginning salary). Notice that the p value for the new managerial variable is also less than .001, which indicates that managers earn a significantly different salary than clerical staff (the reference group for this variable), even after educational level, previous experience and gender are statistically controlled. In contrast, the p value for the new custodial variable is .34, which is greater than .05. This indicates that custodial staff do not earn a significantly different beginning salary than clerical staff (the reference group for this variable), after educational level, previous experience and gender are held constant.

Recall that the degrees of freedom can be found in the ANOVA table in the row labeled "Residual." As shown in that table, the degrees of freedom are equal to 468. Using all of this information, we can now report that the predictors educational level, $b = 664.92$, $SE_b = 96.78$, $\beta = .24$, $t(468) = 6.87$, $p < .001$; previous experience, $b = 10.58$, $SE_b = 2.21$, $\beta = .14$, $t(468) = 4.78$, $p < .001$; gender, $b = -2659.89$, $SE_b = 466.60$, $\beta = -.17$, $t(468) = -5.70$, $p < .001$; and managerial, $b = 12{,}133.70$, $SE_b = 661.83$, $\beta = .59$, $t(468) = 18.33$, $p < .001$; are statistically significant predictors of beginning salary. However, custodial is not a statistically significant predictor of beginning salary, $b = -998.90$, $SE_b = 1{,}035.45$, $\beta = -.03$, $t(468) = -0.96$, $p = .34$.

Note that using the present results, it is not possible to determine whether managerial staff earn a significantly different beginning salary than custodial staff. These results only allow us to determine whether managerial and custodial staff earn a significantly different beginning salary than the reference group of clerical staff. If you wanted to know whether managerial staff earn a significantly different salary than custodial staff, you would need to create a new clerical variable where clerical staff are coded with 1s and managers and custodians are coded with 0s. You would then need to conduct a new regression analysis using managerial and clerical as predictors. By not including custodial as a predictor variable, it will become the new reference group. If the managerial variable was statistically significant in this new analysis, then that would indicate that they earn a significantly different salary than custodial staff. If the managerial variable was not statistically significant in this hypothetical analysis, then that would indicate that managerial staff do not earn a significantly different salary than custodial staff.

Constructing the Equation for the Least-Squares Regression Line

The equation for the least-squares regression line for five predictor variables is: $Y' = b_1X_1 + b_2X_2 + b_3X_3 + b_4X_4 + b_5X_5 + a$. Alternatively, this formula may be expressed as: $\hat{Y} = a + bX_1 + bX_2 + bX_3 + bX_4 + bX_5$. We will label educational level X_1, previous experience X_2, gender X_3, custodial X_4, and managerial X_5.

Coefficients[a]

Model		Unstandardized Coefficients		Standardized Coefficients			Correlations		
		B	Std. Error	Beta	t	Sig.	Zero-order	Partial	Part
1	(Constant)	8809.497	1706.486		5.162	.000			
	Educational Level (years)	664.916	96.781	.244	6.870	.000	.633	.303	.174
	Previous Experience (months)	10.583	2.215	.141	4.778	.000	.045	.216	.121
	GenderCode	-2659.888	466.598	-.168	-5.701	.000	-.457	-.255	-.144
	Custodial	-998.898	1035.451	-.029	-.965	.335	-.061	-.045	-.024
	Managerial	12133.699	661.834	.589	18.333	.000	.782	.647	.464

a. Dependent Variable: Beginning Salary

By looking at the column labeled "B" in the "Coefficients" table, you should be able to see that b_1 (the slope for educational level) is 664.92, b_2 (the slope for previous experience) is 10.58, b_3 (the slope for the gender variable) is $-2,659.89$, b_4 (the slope for custodial) is -998.90, and b_5 (the slope for managerial) is 12,133.70. The intercept, shown at the intersection of the column labeled "B" and the row labeled "(Constant)," is 8,809.50. We now have all of the information we need to construct the equation of our regression line. All we need to do is substitute in the values of b_1, b_2, b_3, and a.

$$Y' = 664.92X_1 + 10.58X_2 - 2,659.89X_3 - 998.90X_4 + 12,133.70X_5 + 8,809.50$$

Using the Regression Equation to Make Predictions

Since one of the predictor variables is not statistically significant, we should not use this equation for the least-squares regression line to make predictions of people's beginning salaries.

Reporting the Results

A sample APA style write-up of these results follows.

A multiple regression analysis was used to predict beginning salary using years of education, months of previous experience, gender, and job category. There were three levels of the job category variable (clerical, custodial, and managerial) and clerical was used as the reference category. The result indicated that a total of 70% of the variability of beginning salary was predicted by the set of predictors ($R^2 = .70$), which was statistically significant, $F(5, 468) = 218.41$ $p < .001$. Years of education, $b = 664.92$, $SE_b = 96.78$, $\beta = .24$, $t(468) = 6.87$, $p < .001$; months of previous experience, $b = 10.58$, $SE_b = 2.21$, $\beta = .14$, $t(468) = 4.78$, $p < .001$; gender, $b = -2,659.89$, $SE_b = 466.60$, $\beta = -.17$, $t(468) = -5.70$, $p < .001$; and managerial, $b = 12,133.70$, $SE_b = 661.83$, $\beta = .59$, $t(468) = 18.33$, $p < .001$; were all found to be significant predictors of beginning salary. Men were coded with 1s and women were coded with 2s, therefore the negative coefficient for gender indicates that women are earning significantly lower beginning salaries than men. Once again, clerical staff were used as the reference group, and therefore the positive coefficient for the managerial variable indicates that managers are earning significantly higher beginning salaries than clerical staff. In contrast, custodial was not a significant predictor of beginning salary, $b = -998.90$, $SE_b = 1,035.45$, $\beta = -.03$, $t(468) = -0.96$, $p = .34$, indicating that custodians do not earn a significantly different beginning salary than clerical staff.

Try On Your Own

The least-squares regression equation can be expanded to include six, seven, or more predictors in the same way that it was expanded to include three, four, or five predictors (by adding in additional b and X terms). Try to build and use a regression equation with six predictors as additional practice. Then, write up the results using APA style. Try using educational level, previous experience, gender, custodial, managerial, and current salary to predict beginning salary.

Advanced Regression

Learning Objectives

In this chapter, you will learn how to conduct hierarchical and stepwise regression analyses. You will also learn how to interpret the results of these analyses and report them using APA style.

HIERARCHICAL REGRESSION

Now that you have mastered simple and multiple regression, you are ready for an introduction to more advanced forms of regression. Hierarchical regression is simply an extension of multiple regression. It allows us to directly compare a simpler regression model with a more complex model, in order to determine whether the addition of a predictor (or a set of predictors) contributes significantly to the regression model. It is commonly used to determine whether a predictor (or set of predictors) can predict a significant amount of variability in the criterion, *over and above* another predictor (or set of predictors). Similarly, it is commonly used to determine whether a predictor variable is significantly related to a criterion variable, after statistically controlling for one or more other predictor variables.

We will once again use the sample dataset "Employee data" for the demonstrations in this chapter. If you saved the data set used for the previous demonstrations, you should use it because we will once again consider the numerically recoded gender variable.

Imagine that you are an industrial organizational consultant who has been hired to assist a group of women in making a case against their company for offering lower beginning salaries to women. The company is countering the women's argument by stating that years of education is the biggest factor in determining the beginning salary offered to each employee and that the reason women tend to earn lower beginning salaries is that they tend to have fewer years of education. In line with their counterargument, you find a significant relationship between gender and years of education. In order to assist these women, you will need to examine whether the gender discrepancy in beginning salaries can be accounted for by differences in education. To explore this possibility, you will need to examine whether adding gender to a regression model predicting beginning salary significantly improves the prediction accuracy over and above (i.e., after controlling for) educational level. Thus, for this analysis beginning salary will be the criterion variable (the Y variable), and years of education and gender will be the predictor variables (the X variables).

Conducting a Hierarchical Regression Analysis (Analyze→Regression→Linear)

To conduct the hierarchical regression analysis, go to **Analyze→Regression→Linear**. A "Linear Regression" dialogue window, like the one shown below, will open. Enter **Beginning Salary** (the criterion variable) into the **Dependent: box** using the corresponding **blue arrow**. Next enter **Educational Level** (the variable you want to control for) into the **Independent(s): box** using the corresponding **blue arrow**.

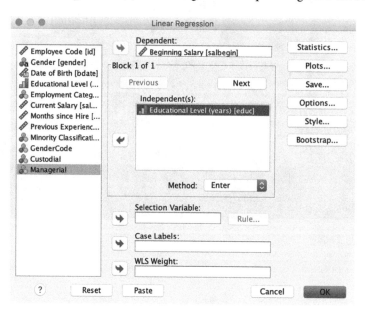

Now, in order to make this a hierarchical regression analysis, you simply need to click on the **Next tab**, located above the Independent(s): box. This will display the second block. Enter **GenderCode** (your primary predictor variable of interest) into the blank **Independent(s): box** that will appear for this second block.

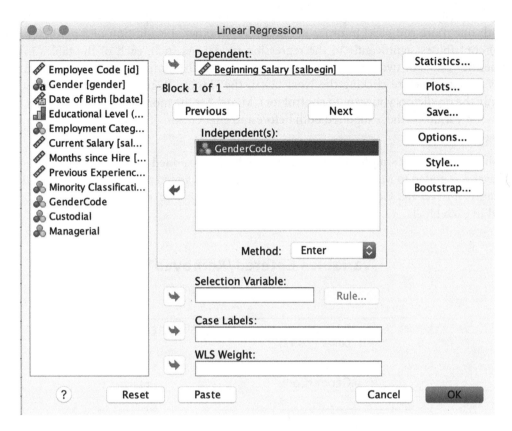

Now click on the **Statistics… tab** in the top right-hand corner of the "Linear Regression" dialogue window. This will open the "Linear Regression: Statistics" dialogue window shown here. **Check** the option for **R squared change**. Finally, click on **Continue,** and then click **OK** on the 'Linear Regression' window to close the windows and execute the analysis.

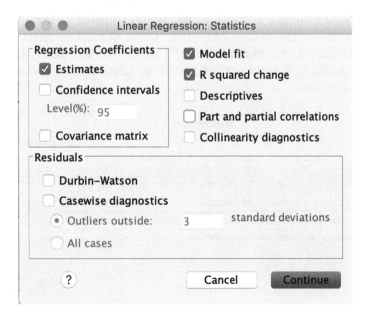

Interpreting the Results

Model Comparisons

As stated previously, hierarchical regression allows us to directly compare two regression models (one simpler with one more complex) in order to determine whether the addition of a predictor (or a set of predictors) contributes significantly to the regression model. As such, each of the tables in the output will include the results of two regression models. Model 1 is the simpler model that contains only the predictor(s) inputted before clicking the Next tab in the "Linear Regression" dialogue window (i.e., it includes only the predictors you want to control for). Model 2 is the more complex model that contains all of the predictor variables (those inputted both before and after clicking the Next tab in the "Linear Regression" dialogue window).

The first table in the output simply displays which variables were added to each model. It shows that Educational Level was added to Model 1 and that GenderCode was added to Model 2. Importantly, however, both Educational Level and GenderCode are included in Model 2. This table only shows which variables were added in each block.

Variables Entered/Removed[a]

Model	Variables Entered	Variables Removed	Method
1	Educational Level (years)[b]	.	Enter
2	GenderCode[b]	.	Enter

a. Dependent Variable: Beginning Salary

b. All requested variables entered.

Next, we will examine the "Model Summary" table shown below. It contains the primary results of interest. The results shown in the row labeled "1" are for a simple regression analysis using only educational level to predict beginning salary. As such, they are redundant to those computed for the simple regression analysis demonstrated in Chapter Four. The column labeled "R" shows that the correlation between educational level and beginning salary is .63, the column labeled "R Square" shows that coefficient of determination is .40, and the column labeled "Std. Error of the Estimate" indicates that the standard error of estimate is $6,098.26.

Model Summary

Model	R	R Square	Adjusted R Square	Std. Error of the Estimate	R Square Change	F Change	df1	df2	Sig. F Change
1	.633[a]	.401	.400	$6,098.259	.401	315.897	1	472	.000
2	.680[b]	.462	.460	$5,784.256	.061	53.637	1	471	.000

a. Predictors: (Constant), Educational Level (years)

b. Predictors: (Constant), Educational Level (years), GenderCode

Results pertaining to whether or not Model 1 is statistically significant can be found both in the ANOVA table shown on the following page and in the "Change Statistics" section of the "Model Summary" table shown previously. Despite the name of this section, since for the first model there is no change to consider, the values displayed in the top row pertain to the significance of the model as a whole.[1] The "F Change" column provides the F statistic, the columns "df1" and "df2" provide both degrees of freedom values, and the "Sig. F Change" column displays the p value for the F statistic. By reviewing these statistics, you should be able to determine that the first model is statistically significant, $r^2 = .40$, $F(1, 472) = 315.90$, $p < .001$. In other words, educational level predicts a significant portion of variance in beginning salary.

The results displayed in the row labeled "2" are for the second, more complex, model that includes both educational level and gender as predictors. First, by reviewing the left side of the table, you should be able to see that the value of the multiple correlation is .68, the value of the multiple coefficient of determination (R^2) is .46, and the standard error of estimate is \$5,784.26.

For the second, more complex, model, the term "Change Statistics" is appropriate because the statistics displayed on the right side of the "Model Summary" table concern the change in the R^2 value that resulted from adding gender to the model. The column labeled "R Square Change" shows that adding the variable gender increased the R^2 value of the model by .06. You can confirm this for yourself by comparing the two R^2 values displayed in the "R Square" column ($.462 - .401 = .061$). The column labeled "F Change" shows that the F statistic for this change in R^2 is equal to 53.64. The columns labeled "df1" and "df2" display the degrees of freedom for this F change statistic. Finally, you should be able to see that this change is statistically significant because the p value associated with the F change statistic, displayed in the last column, is less than .001. Change statistics are symbolized using a Δ. Therefore, these results would be reported as $\Delta R^2 = .06$, $\Delta F(1, 471) = 53.64$, $p < .001$. These results indicate that gender contributes a significant portion of unique variance in the prediction of beginning salary, above and beyond what educational level can predict. In other words, there is a relationship between gender and beginning salary that is independent of educational level. It appears these women do have a case against this company because differences in years of education do not account for the gender discrepancy in beginning salary; rather, women are paid significantly lower beginning salaries even after differences in their years of education are taken into consideration.

Model Statistics

The "ANOVA" table, shown here, simply provides the degrees of freedom, F statistics, and p values for the two regression models. The results of the first model are displayed in the upper portion of the table labeled Model 1. As you can once again see, the simpler model that includes only educational level as a predictor is statistically significant, $R^2 = .40$, $F(1, 472) = 315.90$, $p < .001$. As you should expect then, the more complex model that includes both educational level and gender as predictors should also be statistically significant. And indeed, as shown in the lower portion of the table labeled Model 2, the two predictor model is also statistically significant, $R^2 = .46$, $F(2, 471) = 202.38$, $p < .001$. However, these results are not terribly interesting, and, in many cases, we would not report them. For hierarchical regression, we focus more on the change statistics described earlier.

---

[1] You can confirm this for yourself by comparing the values provided in the top section of the "Model Summary" table with those provided in the upper portion of the "ANOVA" table and by comparing the R Square and R Square Change statistics provided in the "Model Summary" table.

ANOVA[a]

Model		Sum of Squares	df	Mean Square	F	Sig.
1	Regression	1.175E+10	1	1.175E+10	315.897	.000[b]
	Residual	1.755E+10	472	37188762.8		
	Total	2.930E+10	473			
2	Regression	1.354E+10	2	6.771E+9	202.381	.000[c]
	Residual	1.576E+10	471	33457613.3		
	Total	2.930E+10	473			

a. Dependent Variable: Beginning Salary

b. Predictors: (Constant), Educational Level (years)

c. Predictors: (Constant), Educational Level (years), GenderCode

Predictor Statistics

Finally, the coefficients table, shown below, displays the regression coefficients (in both unstandardized and standardized form), the t statistics, and p values for each of the predictors, separately for each of the regression models. Once again, the statistics for the simpler (one predictor) model are presented in the upper portion of the table (the portion labeled "1"). You should be able to see that educational level is a significant predictor of beginning salary. And from Chapters Four and Five, you should also know that these results would be reported as: $b = 1{,}727.53$, $SE_b = 97.20$, $\beta = .63$, $t(472) = 17.77$, $p < .001$. Once again, the value of the degrees of freedom that need to be reported with the t statistic can be found in the "Model Summary" table in the column labeled "df2."

The statistics for the more complex (two predictor) model are presented in the lower portion of the table (the portion labeled "2"). By referring to this portion of the table, you should be able to see that both educational level, $b = 1{,}470.32$, $SE_b = 98.66$, $\beta = .54$, $t(471) = 14.90$, $p < .001$, and gender, $b = -4{,}180.77$, $SE_b = 570.85$, $\beta = -.26$, $t(471) = -7.32$, $p < .001$, are significant predictors. Moreover, the slope of the gender variable is inverse, confirming that being a woman is associated with being paid lower beginning salaries. Indeed, this result can be interpreted to mean that women earn beginning salaries that are, on average, \$4,180.77 less than men, even after controlling for differences in years of education.

Coefficients[a]

Model		Unstandardized Coefficients		Standardized Coefficients		
		B	Std. Error	Beta	t	Sig.
1	(Constant)	−6290.967	1340.920		−4.692	.000
	Educational Level (years)	1727.528	97.197	.633	17.773	.000
2	(Constant)	3265.086	1822.140		1.792	.074
	Educational Level (years)	1470.321	98.655	.539	14.904	.000
	GenderCode	−4180.769	570.853	−.265	−7.324	.000

a. Dependent Variable: Beginning Salary

Excluded Variables

The last table provided in the output, labeled "Excluded Variables" is not very relevant or meaningful. It simply displays what the beta weight (standardized slope), t statistic, and p value associated with the gender variable would be if that variable was included in Model 1. Not surprisingly, the values are redundant with those for Model 2, which included gender. The table also displays the partial correlation for the gender variable, which represents the correlation between gender and beginning salary, with the effects of education on both gender and beginning salary statistically controlled.

Finally, the last column of the table shows a collinearity statistic called tolerance, which represents the proportion of variance in the predictor of interest (in this case gender) that is unique from the other predictor(s) included in Model 1 (in this case educational level). The value of .87 reported in the table indicates that 87.33% of the variance in gender is unique from (i.e., is unrelated to) educational level. This value is high and indicates that it would be appropriate to include gender as a predictor variable in models with educational level. In contrast, low tolerance values, such as values between .10 and .20, might be a cause for concern, as they would indicate that only 10% to 20% of the variance in the predictor is unique from other predictors in the model. Since only unique variance is considered when determining whether a predictor is statistically significant, variables with low tolerance statistics would be unlikely to be significant predictors.

Excluded Variables[a]

Model		Beta In	t	Sig.	Partial Correlation	Collinearity Statistics Tolerance
1	GenderCode	−.265[b]	−7.324	.000	−.320	.873

a. Dependent Variable: Beginning Salary

b. Predictors in the Model: (Constant), Educational Level (years)

Reporting the Results

We could report the following results to the company:

A hierarchical regression analysis was conducted, using gender to predict beginning salary, after statistically controlling for years of education. The results showed that the inclusion of gender significantly improved the prediction accuracy of the model, $\Delta R^2 = .06$, $\Delta F(1, 471) = 53.64$, $p < .001$. Years of education was a significant predictor of beginning salary, $b = 1,470.32$, $SE_b = 98.66$, $\beta = .54$, $t(471) = 14.90$, $p < .001$, after controlling for gender. Moreover, gender was a significant predictor of beginning salary, after controlling for years of education, $b = -4,180.77$, $SE_b = 570.85$, $\beta = -.26$, $t(471) = -7.32$, $p < .001$. These results clearly demonstrate that there is a relationship between gender and beginning salary that is independent of education. More specifically, they indicate that women are paid $4,180.77 less than men, on average, even after potential differences in the years of education of these two groups are taken into consideration.

Conducting a More Complex Hierarchical Regression Analysis (Analyze→Regression→Linear)

In the previous example, our simpler model contained only one predictor variable, and our more complex model contained only two predictor variables. But hierarchical regression can be used to examine models that are far more complex. For the next demonstration, we will include two predictor variables in the simpler model and two predictor variables in the more complex model.

Imagine that the company under investigation disputed the results you presented to them and claimed that women also have fewer months of previous experience, and that, combined with fewer years of education, accounts for the gender discrepancy in beginning salary. At the same time, a broader diversity advocates group heard about your case and, based on their findings that minorities also earn significantly lower beginning salaries, they asked to join the case against this company. Thus, for this more complex hierarchical regression analysis, you will need to examine whether gender and minority status significantly predict beginning salary after controlling for educational level and previous experience. This will allow you to determine whether lower levels of education and fewer months of previous experience account for the lower beginning salaries offered to women and minorities at this company.

To run the analysis, go to **Analyze→Regression→Linear**. A "Linear Regression" dialogue window, like the one shown below, will open. Beginning Salary (the criterion variable) should still appear in the Dependent: box from the previous analysis. If it doesn't, then move it into this box using the corresponding blue arrow. The dialogue window will likely open, displaying Block 2 of 2 with GenderCode displayed from the previous analysis. Click **Previous** to display the first block. Educational level should still appear in the first block (the first Independent(s): box). If it doesn't, then move it into this box using the corresponding blue arrow. Add **Previous Experience** to the first **Independent(s): box** with Educational Level using the corresponding **blue arrow**. Your dialogue window should now look like the one displayed here.

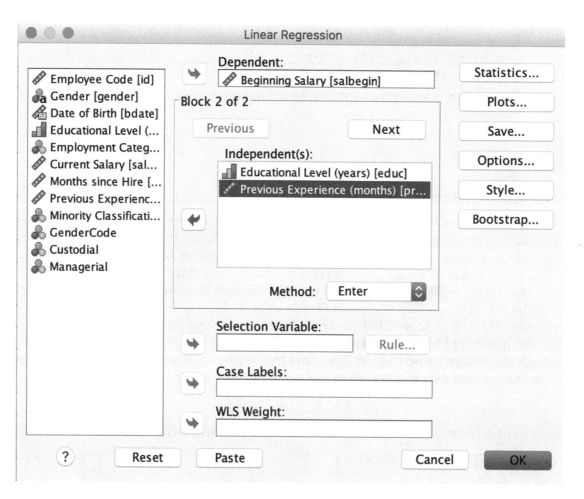

Click on the **Next tab**. GenderCode should still appear in this block. If it doesn't, then add it. Next add **Minority Classification** into the **Independent(s) box** with GenderCode. Your "Linear Regression" dialogue window should now look like the one displayed here.

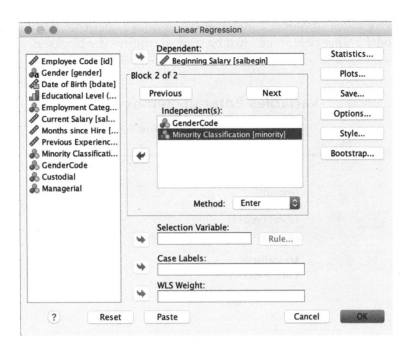

Next click on the **Statistics... tab**. This will open the "Linear Regression: Statistics" dialogue window shown here. Make sure the option for R squared change is still checked. Click on **Continue** and then **OK** to close the dialogue windows and execute the analysis.

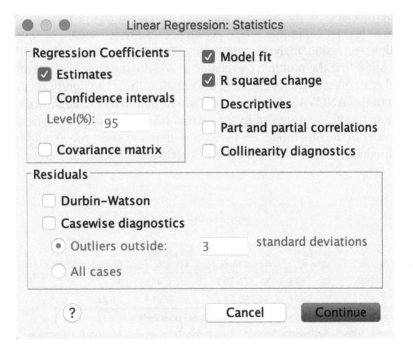

Interpreting the Results
Model Comparisons

The first table in the output displays the predictor variables that were added to each Model. They show that Previous Experience and Educational Level were added to Model 1 and that Minority Classification and GenderCode were added to Model 2. Remember, Previous Experience and Educational Level are also included in Model 2. This table should just be used to confirm the blocks in which each predictor was added.

Variables Entered/Removed[a]

Model	Variables Entered	Variables Removed	Method
1	Previous Experience (months), Educational Level (years)[b]	.	Enter
2	Minority Classification, GenderCode[b]	.	Enter

a. Dependent Variable: Beginning Salary

b. All requested variables entered.

Once again, the "Model Summary" table shown below contains the primary results of interest. The results shown in the row labeled "1" are for a multiple regression analysis using educational level and previous experience to predict beginning salary. As such, they are redundant to those computed for the two-predictor multiple regression analysis reviewed in Chapter Five. Consistent with the results of that analysis, the results show that the multiple correlation is .67, the multiple coefficient of determination is .45 and the standard error of estimate is $5,871.76. Information about the significance of Model 1 can once again be found in both the "ANOVA" table and in the "Model Summary" table shown below. In the "Model Summary" table, these results are provided in the section labeled "Change Statistics," even though there is no change to consider in the first model. By reviewing these statistics, you should be able to determine that the model with educational level and previous experience is statistically significant, $R^2 = .45$, $F(2, 471) = 189.43$, $p < .001$.

Model Summary

Model	R	R Square	Adjusted R Square	Std. Error of the Estimate	Change Statistics				
					R Square Change	F Change	df1	df2	Sig. F Change
1	.668[a]	.446	.443	$5,871.763	.446	189.427	2	471	.000
2	.706[b]	.499	.494	$5,596.641	.053	24.723	2	469	.000

a. Predictors: (Constant), Previous Experience (months), Educational Level (years)

b. Predictors: (Constant), Previous Experience (months), Educational Level (years), Minority Classification, GenderCode

The results displayed in the row labeled "2" are for the second, more complex, model that includes educational level, previous experience, gender, and minority classification as predictors. First, by reviewing the left side of the table, you should be able to see that for this more complex model, the value of the multiple correlation is .71, the value of the multiple coefficient of determination is .50, and the standard error of estimate is $5,596.64. The right side of the "Model Summary" table shows that adding gender and minority classification to the regression model increased the R^2 value by .05 and the F statistic for this change in R^2 is equal to 24.72. Finally, you should be able to see that this change is statistically significant because the associated p value is less than .001. These results would be reported as $\Delta R^2 = .05$, $\Delta F(2, 469) = 24.72$, $p < .001$.

These results indicate that, as a set, gender and minority classification account for a significant proportion variability in beginning salary, over and above that accounted for by educational level and previous experience. In other words, gender and minority classification together show a significant relationship with beginning salary after education and previous experience are statistically controlled. It appears that fewer years of experience and fewer months of previous experience do not account for the diminished salaries offered to these groups. Our case against this company has been strengthened.

Predictor Statistics

Since we added more than one predictor variable to the second step of the hierarchical regression analysis, we need to examine the significance of each of the predictors before we can reach too strong a conclusion. Let's skip over the "ANOVA" table for a minute and jump right to the "Coefficients" table shown here. Once again, the statistics for the simpler (two-predictor) model are presented in the upper portion of the table (labeled Model 1). Consistent with the two-predictor multiple regression analysis we conducted in Chapter Five, the results show that educational level, $b = 18,788.21$, $SE_b = 96.72$, $\beta = .69$, $t(471) = 19.42$, $p < .001$, and previous experience, $b = 16.47$, $SE_b = 2.67$, $\beta = .22$, $t(471) = 6.17$, $p < .001$, are each significant predictors of beginning salary. Note that the degrees of freedom that need to be reported with these predictor statistics can be found in the "Model Summary" table in the column labeled "df2" and the row labeled "Model 1."

Coefficients[a]

Model		Unstandardized Coefficients B	Unstandardized Coefficients Std. Error	Standardized Coefficients Beta	t	Sig.
1	(Constant)	–9902.786	1417.474		–6.986	.000
	Educational Level (years)	1878.211	96.717	.688	19.420	.000
	Previous Experience (months)	16.470	2.668	.219	6.174	.000
2	(Constant)	405.585	2041.776		.199	.843
	Educational Level (years)	1574.258	102.343	.577	15.382	.000
	Previous Experience (months)	12.812	2.659	.170	4.818	.000
	GenderCode	–3670.725	578.845	–.233	–6.341	.000
	Minority Classification	–2339.717	634.479	–.123	–3.688	.000

a. Dependent Variable: Beginning Salary

The statistics for the more complex (four-predictor) model are presented in the lower portion of the table labeled "Model 2." As displayed in this portion of the table, both gender, $b = -3,670.73$, $SE_b = 578.85$, $\beta = -.23$, $t(469) = -6.34$, $p < .001$, and minority classification, $b = -2,339.72$, $SE_b = 634.48$, $\beta = -.12$, $t(469) = -3.69$, $p < .001$, are significant predictors of beginning salary. Once again, the slope for each of these predictors is inverse, confirming that being a woman (coded with a higher value) is associated with lower beginning salaries and being a minority (also coded with a higher value) is associated with lower beginning salaries. Moreover, by directly comparing the beta weights for gender ($\beta = -.23$) and minority classification ($\beta = -.12$), we can determine that after controlling for educational level and previous experience, the relationship between gender and beginning salary is stronger than the relationship between minority status and beginning salary. By examining the unstandardized regression coefficients, we can further determine that we would predict a woman in this company to be paid a beginning salary that is $3,670.73 lower than the beginning salary of a man with the same number of years of education, months of previous experience, and minority classification. Similarly, we would predict a minority to be paid a beginning salary that is $2,339.72 lower than a nonminority of the same gender, with the same number of years of education, and the same number of months of previous experience. Apparently, this company is both sexist and racist, but they are a bit more sexist than racist.

Note that if one of the predictors demonstrated a significant bivariate correlation with the criterion but then was not statistically significant when entered into the second block (Model 2), it would suggest that the relationship between the nonsignificant predictor and the criterion could be explained by other predictors entered into the model. As a hypothetical example, assume we found that the R^2 change and F change statistics were statistically significant (that adding the set of predictors significantly improved the prediction accuracy of the model), that the predictor gender was statistically significant, but that the predictor minority classification was *not* significant. This pattern of findings would suggest that while years of education and months of previous experience do not account for the gender discrepancy in beginning salary, years of education and previous experience *do* account for the decreased beginning salaries offered to minorities in this company.

Model Statistics

Although we already have all of the information we need, we will briefly consider the "ANOVA" table displayed below. This table simply provides the degrees of freedom, F statistics, and p values for both of the regression models. As you can once again see, the simpler model that includes educational level and previous experience as predictors is statistically significant, $R^2 = .45$, $F(2, 471) = 189.43$, $p < .001$. Accordingly, the more complex model that includes all four predictor variables is also statistically significant, $R^2 = .50$, $F(4, 469) = 116.62$, $p < .001$. Once again, these results are not terribly interesting, and, in most cases, we would not report them.

ANOVA[a]

Model		Sum of Squares	df	Mean Square	F	Sig.
1	Regression	1.306E+10	2	6.531E+9	189.427	.000[b]
	Residual	1.624E+10	471	34477599.6		
	Total	2.930E+10	473			
2	Regression	1.461E+10	4	3.653E+9	116.615	.000[c]
	Residual	1.469E+10	469	31322390.6		
	Total	2.930E+10	473			

a. Dependent Variable: Beginning Salary

b. Predictors: (Constant), Previous Experience (months), Educational Level (years)

c. Predictors: (Constant), Previous Experience (months), Educational Level (years), Minority Classification, GenderCode

Excluded Variables

The last table provided in the output, labeled "Excluded Variables," is also not very relevant or meaningful. It simply displays what the beta weight [standardized slope (β)], t statistic, and p value associated with the gender and minority classification variables would be if each were included in Model 1 without the other. In other words, the "Beta In" value reported for gender indicates that if we included gender in a model with educational level and previous experience (but not minority classification), then its beta weight would be –.22. Similarly, the "Beta In" value reported for minority classification indicates that if we included minority classification in a model with educational level and previous experience (but not gender), then its beta weight would be –.10.

Excluded Variables[a]

Model		Beta In	t	Sig.	Partial Correlation	Collinearity Statistics Tolerance
1	GenderCode	–.218[b]	–5.909	.000	–.263	.804
	Minority Classification	–.101[b]	–2.919	.004	–.133	.969

a. Dependent Variable: Beginning Salary

b. Predictors in the Model: (Constant), Previous Experience (months), Educational Level (years)

The table also displays the partial correlation for the gender variable (–.26), which represents the correlation between gender and beginning salary, with the effects of education and previous experience on both gender and beginning salary statistically controlled. Similarly, the partial correlation of –.13 indicates that the correlation between minority classification and beginning salary is –.13, after the influence of education and previous experience on both minority classification and beginning salary have been statistically controlled.

Finally, the last column of the table shows a collinearity statistic called tolerance, which represents the proportion of variance in the predictor of interest that is unique from the other predictors included in Model 1. The value of .80 reported in the table for gender indicates that 80.40% of the variance in gender is unique from (i.e., is unrelated to) educational level and previous experience. The value of .96 provided for minority classification indicates that 96.91% of the variability in minority classification is unique from (i.e., does not overlap with) educational level and previous experience. Once again, these values are high and indicate that it is appropriate to include these predictor variables in models with educational level and previous experience.

Reporting the Results

We could report the following results to the company:

A hierarchical regression analysis was conducted, using gender and minority classification to predict beginning salary after entering years of education and months of previous experience. The results showed that the inclusion of gender and minority classification significantly improved the prediction accuracy of the regression model, $\Delta R^2 = .05$, $\Delta F(2, 469) = 24.72$, $p < .001$. Moreover, both gender, $b = -3670.73$, $SE_b = 578.85$, $\beta = -.23$, $t(469) = -6.34$, $p < .001$, and minority classification, $b = -2,339.72$, $SE_b = 634.48$, $\beta = -.12$, $t(469) = -3.69$, $p < .001$, were significant predictors of beginning salary, after controlling for years of educational and previous experience. These results clearly show that there are relationships between beginning salary and both gender and minority classification that are independent of years of education and months of previous experience. In other words, women and minorities are paid significantly lower beginning salaries even after potential differences in their years of education and previous experience are taken into consideration. More specifically, on average, women earn beginning salaries that are $3,670.73 less than those earned by men with comparable education, experience, and minority status. Minorities earn beginning salaries that are, on average, $2,339.72 less than nonminorities with comparable education, experience, and gender.

STEPWISE REGRESSION

While hierarchical regression is used to address theoretical questions about the nature of relationships between variables, stepwise regression is an entirely exploratory technique. Stepwise regression is empirically, rather than theoretically, based. Researchers using stepwise regression enter all variables of interest into the regression model, and SPSS finds the model that offers the best predictive utility. It begins by entering the predictor variable that has the strongest relationship to the criterion, and it continues to enter predictor variables, one at a time, in the order of their strength of relationship to the criterion (in the order of the magnitude of their semipartial correlations). It stops once the addition of predictors ceases to significantly improve the predictive accuracy of the model (until the addition of more predictors ceases to increase R^2 by a significant amount).

Stepwise regression is not a highly respected technique because, as scientists, we typically prefer models that are built on the basis of theories. With stepwise regression, a mindless computer program is building the model. Moreover, we typically build models using the data from a sample with the ultimate goal of generalizing to the population. Stepwise regression is often said to capitalize on chance because the decisions about which variables to include are made on the basis of potentially trivial differences in semipartial correlations computed from a single sample, even though some variability in the statistics from sample to sample is expected. It is often said to "overfit" the data because the model derived from a single sample is too close to the sample and may not generalize well to the population. For this reason, the models often cannot be successfully replicated with different samples from the same population. We are less likely to capitalize on chance when we use a theory to make decisions about which predictors to include and which models to test. The bottom line is that it is not very respectable to have a computer program making your decisions for you.

Nevertheless, if we are simply interested in exploring relationships between variables and finding the model that offers the best predictive utility, stepwise regression can be a useful technique. Using the "Employee data" data set with the numerically recoded gender variable, we will attempt to find the model that best predicts *current* salary (note we are no longer considering beginning salary).

Conducting a Stepwise Regression Analysis (Analyze→Regression→Linear)

To run the analysis, go to **Analyze→Regression→Linear**. A "Linear Regression" dialogue window will open. Make sure that any variables that were entered for the previous analysis are cleared out of the Dependent and Independent(s) boxes. (Be sure to click the Previous tab to clear out variables entered into the first block of the hierarchical regression analysis.) Enter **Current Salary** (the criterion variable) into the **Dependent: box** using the corresponding **blue arrow**. Next enter **Educational Level, Beginning Salary, Months since Hire, Previous Experience, Minority Classification,** and **GenderCode** into the **Independent(s): box** using the corresponding **blue arrow**. Note that we will not enter Employee Code or Date of Birth because it would not make any sense whatsoever to include either of those variables as predictors (we don't want to be entirely mindless here). Also, for the sake of simplicity we will not enter

Employment Category. Next click on the tab labeled **Method** and select **Stepwise.** Next click on the **Statistics...** tab.

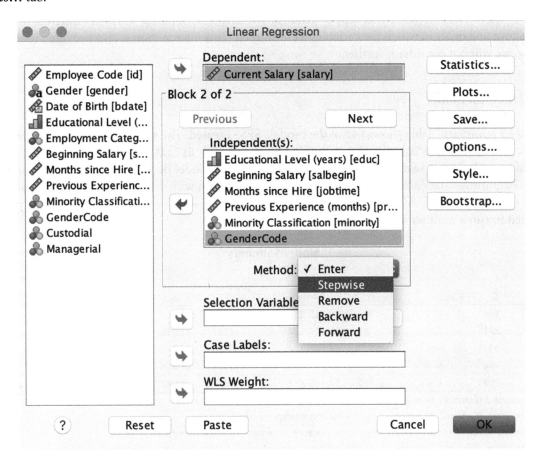

Clicking the Statistics tab will open the "Linear Regression: Statistics" dialogue window shown below. Check the option for **R squared change.** Click **Continue,** and then click **OK** on the main "Linear Regression" window to close the windows and execute the analysis.

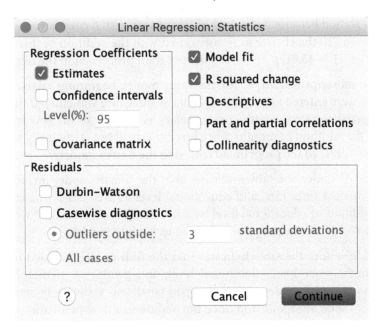

Interpreting the Results

Several tables will now appear in the output window. The first table, labeled "Variables Entered/Removed" simply shows the order in which the variables were entered. This information is also provided in the other tables, so we will not consider it further.

Finding the Best Model

The "Model Summary" table shows five of the models SPSS created. The subscript under the table labeled "a" shows that Model 1 was built using a simple regression analysis that included only beginning salary as a predictor of current salary. The program started with this model because, of all of the predictors we entered, beginning salary shows the highest bivariate correlation with current salary. Since, as shown in the first row, that model was statistically significant, $r^2 = .77$, $F(1, 472) = 1,622.12$, $p < .001$, the program proceeded to run a multiple regression analysis.

Model Summary

Model	R	R Square	Adjusted R Square	Std. Error of the Estimate	R Square Change	F Change	df1	df2	Sig. F Change
					Change Statistics				
1	.880ª	.775	.774	$8,115.356	.775	1622.118	1	472	.000
2	.891ᵇ	.793	.793	$7,776.652	.019	43.010	1	471	.000
3	.897ᶜ	.804	.803	$7,586.187	.010	24.948	1	470	.000
4	.900ᵈ	.810	.809	$7,465.139	.007	16.366	1	469	.000
5	.902ᵉ	.814	.812	$7,410.457	.003	7.947	1	468	.005

a. Predictors: (Constant), Beginning Salary

b. Predictors: (Constant), Beginning Salary, Previous Experience (months)

c. Predictors: (Constant), Beginning Salary, Previous Experience (months), Months since Hire

d. Predictors: (Constant), Beginning Salary, Previous Experience (months), Months since Hire, Educational Level (years)

e. Predictors: (Constant), Beginning Salary, Previous Experience (months), Months since Hire, Educational Level (years), GenderCode

As shown under the table next to the subscript labeled "b," the first multiple regression analysis included beginning salary and previous experience as predictors of current salary. Previous experience was selected by SPSS to be entered second because it shows the highest semipartial correlation with current salary. As shown in the row labeled "2," the change in R^2 associated with the addition of this predictor was significant, $\Delta R^2 = .02$, $\Delta F(1, 471) = 43.01$, $p < .001$, so the program proceeded to the next step.

As shown next to the subscript labeled "c" for the third model, beginning salary, previous experience, and months since hire were entered as predictors. Months since hire was entered third because it has the second largest semipartial correlation with current salary. As shown in the row labeled "3," the change in R^2 associated with the addition of months since hire was small but statistically significant, $\Delta R^2 = .01$, $\Delta F(1, 470) = 24.95$, $p < .001$, so the program advanced to the fourth model.

The subscript labeled "d" under the table indicates that the fourth model included beginning salary, previous experience, months since hire, and educational level as predictors. The change in the R^2 value associated with the addition of educational level was also small but statistically significant, $\Delta R^2 = .007$, $\Delta F(1, 469) = 16.37$, $p < .001$, so the program advanced to the fifth model.

The subscript labeled "e" under the table indicates that the fifth model included beginning salary, previous experience, months since hire, educational level, and gender as predictors. The change in R^2 associated with the addition of gender was once again small but statistically significant, $\Delta R^2 = .003$, $\Delta F(1, 468) = 7.95$, $p = .005$. SPSS will stop once the addition of new predictors ceases to improve the model. As such, this last model is the one that SPSS has identified as being the best model to predict current salary.

Model Statistics

The "ANOVA" table shown here provides the degrees of freedom, F statistics, and p values for all five of the regression models. Since stepwise regression is used to find the model that best fits the data, we are only interested in the statistics provided in the portion of the table associated with Model 5, which we just identified as the best model. As you can see, the fifth regression model that includes beginning salary, previous experience, months since hire, educational level, and gender as predictors, is statistically significant, $F(5, 468) = 408.69, p < .001$.

ANOVA[a]

Model		Sum of Squares	df	Mean Square	F	Sig.
1	Regression	1.068E+11	1	1.068E+11	1622.118	.000[b]
	Residual	3.109E+10	472	65858997.2		
	Total	1.379E+11	473			
2	Regression	1.094E+11	2	5.472E+10	904.752	.000[c]
	Residual	2.848E+10	471	60476323.3		
	Total	1.379E+11	473			
3	Regression	1.109E+11	3	3.696E+10	642.151	.000[d]
	Residual	2.705E+10	470	57550239.5		
	Total	1.379E+11	473			
4	Regression	1.118E+11	4	2.794E+10	501.450	.000[e]
	Residual	2.614E+10	469	55728306.8		
	Total	1.379E+11	473			
5	Regression	1.122E+11	5	2.244E+10	408.692	.000[f]
	Residual	2.570E+10	468	54914875.0		
	Total	1.379E+11	473			

a. Dependent Variable: Current Salary

b. Predictors: (Constant), Beginning Salary

c. Predictors: (Constant), Beginning Salary, Previous Experience (months)

d. Predictors: (Constant), Beginning Salary, Previous Experience (months), Months since Hire

e. Predictors: (Constant), Beginning Salary, Previous Experience (months), Months since Hire, Educational Level (years)

f. Predictors: (Constant), Beginning Salary, Previous Experience (months), Months since Hire, Educational Level (years), GenderCode

Predictor Statistics

Finally, we will consider the "Coefficients" table shown next since it contains our regression coefficients and information on each of the predictors. Once again, we are interested primarily in the statistics provided in the portion of the table pertaining to Model 5, since this model was identified as being the best model to predict current salary. We can use the information in this portion of this table to report the relevant statistics for each of our predictors and to construct the equation for the least-squares regression line. Note, however, that the value of the degrees of freedom that we will need to report for each predictor need to be found in the "Model Summary" table in the column labeled "df2" or in the ANOVA table in the row labeled "Residual." These results show that beginning salary, $b = 1.72$, $SE_b = 0.06$, $\beta = .79$, $t(468) = 28.47$, $p < .001$, previous experience, $b = -19.44$, $SE_b = 3.58$, $\beta = -.12$, $t(468) = -5.42$, $p < .001$, months since hire, $b = 154.54$, $SE_b = 34.08$, $\beta = .09$, $t(468) = 4.53$, $p < .001$, educational level, $b = 593.03$, $SE_b = 166.63$, $\beta = .10$, $t(468) = 3.56$, $p < .001$, and gender, $b = -2,232.92$, $SE_b = 792.08$, $\beta = -.07$, $t(468) = -2.82$, $p = .005$, are all significant predictors of current salary.

Coefficients[a]

Model		Unstandardized Coefficients B	Std. Error	Standardized Coefficients Beta	t	Sig.
1	(Constant)	1928.206	888.680		2.170	.031
	Beginning Salary	1.909	.047	.880	40.276	.000
2	(Constant)	3850.718	900.633		4.276	.000
	Beginning Salary	1.923	.045	.886	42.283	.000
	Previous Experience (months)	-22.445	3.422	-.137	-6.558	.000
3	(Constant)	-10266.629	2959.838		-3.469	.001
	Beginning Salary	1.927	.044	.888	43.435	.000
	Previous Experience (months)	-22.509	3.339	-.138	-6.742	.000
	Months since Hire	173.203	34.677	.102	4.995	.000
4	(Constant)	-16149.671	3255.470		-4.961	.000
	Beginning Salary	1.768	.059	.815	30.111	.000
	Previous Experience (months)	-17.303	3.528	-.106	-4.904	.000
	Months since Hire	161.486	34.246	.095	4.715	.000
	Educational Level (years)	669.914	165.596	.113	4.045	.000
5	(Constant)	-10317.115	3837.191		-2.689	.007
	Beginning Salary	1.723	.061	.794	28.472	.000
	Previous Experience (months)	-19.436	3.583	-.119	-5.424	.000
	Months since Hire	154.536	34.085	.091	4.534	.000
	Educational Level (years)	593.031	166.630	.100	3.559	.000
	GenderCode	-2232.917	792.078	-.065	-2.819	.005

a. Dependent Variable: Current Salary

Excluded Variables

The last table, labeled "Excluded Variables," shows the variables that were excluded from the five regression models. As further described in the section on "Hierarchical Regression," this table simply shows what the beta weights, t statistics, p values, and partial correlations for each of these predictors would have been if each had been included in each model with the other predictors included in that model. For instance, the top row, labeled "Model 1" "Educational Level (years)" indicates that if Educational Level was included as a predictor variable in Model 1 along with beginning salary (and beginning salary alone), then its beta weight would have been .17, the corresponding t statistic would be 6.36, and the predictor would be significant with $p < .001$. It also indicates that the partial correlation between years of education and current salary, with the influence of beginning salary on education and current salary statistically controlled, is .28. Finally, it displays a tolerance statistic of .60, which indicates that 59.91% of the variability in educational level is unique from (i.e., does not overlap with) beginning salary.

Excluded Variables[a]

Model		Beta In	t	Sig.	Partial Correlation	Collinearity Statistics Tolerance
1	Educational Level (years)	.172[b]	6.356	.000	.281	.599
	Months since Hire	.102[b]	4.750	.000	.214	1.000
	Previous Experience (months)	-.137[b]	-6.558	.000	-.289	.998
	Minority Classification	-.040[b]	-1.794	.073	-.082	.975
	GenderCode	-.061[b]	-2.482	.013	-.114	.791
2	Educational Level (years)	.124[c]	4.363	.000	.197	.520
	Months since Hire	.102[c]	4.995	.000	.225	1.000
	Minority Classification	-.019[c]	-.869	.385	-.040	.952
	GenderCode	-.088[c]	-3.740	.000	-.170	.771
3	Educational Level (years)	.113[d]	4.045	.000	.184	.516
	Minority Classification	-.024[d]	-1.127	.260	-.052	.950
	GenderCode	-.079[d]	-3.406	.001	-.155	.765
4	Minority Classification	-.024[e]	-1.185	.237	-.055	.950
	GenderCode	-.065[e]	-2.819	.005	-.129	.744
5	Minority Classification	-.033[f]	-1.618	.106	-.075	.930

a. Dependent Variable: Current Salary

b. Predictors in the Model: (Constant), Beginning Salary

c. Predictors in the Model: (Constant), Beginning Salary, Previous Experience (months)

d. Predictors in the Model: (Constant), Beginning Salary, Previous Experience (months), Months since Hire

e. Predictors in the Model: (Constant), Beginning Salary, Previous Experience (months), Months since Hire, Educational Level (years)

f. Predictors in the Model: (Constant), Beginning Salary, Previous Experience (months), Months since Hire, Educational Level (years), GenderCode

Once again, the bottom portion of the table, labeled "Model 5," is of primary interest because it shows that if minority classification (the only variable that was excluded from the model) had been included in the model, it would not have been a significant predictor of current salary, $\beta = -.03$, $t(467) = -1.62$, $p = .11$, and, accordingly, would not have increased R^2 by a significant amount. This confirms that the model containing only the five remaining predictors is the most superior model to predict current salary.

Constructing the Equation for the Least-Squares Regression Line

The values listed in the "Coefficients" table shown previously can also be used to construct the least-squares regression line. The equation for a model with five predictor variables is: $Y' = b_1X_1 + b_2X_2 + b_3X_3 + b_4X_4 + b_5X_5 + a$. Alternatively, this formula may be expressed as: $\hat{Y} = a + bX_1 + bX_2 + bX_3 + bX_4 + bX_5$.

Once again, we are interested only in the bottom portion of the table pertaining to Model 5. As reviewed in Chapters Four and Five, all of the regression coefficients that we need to construct the equation for the least-squares regression line can be found in the column labeled "B." By referring to the rows labeled with the variable names, you should be able to determine that the value of b_1 (the slope for beginning salary) is 1.72, the value of b_2 (the slope for previous experience) is -19.44, the value of b_3 (the slope for months since hire) is 154.54, the value of b_4 (the slope for educational level) is 593.03, and the slope of b_5 (the slope for gender) is $-2,232.92$. Finally, by referring to the row labeled "(Constant)," you should be able to see that the value of a, is $-10,317.12$. Therefore, the equation for the least-squares regression line is:

$$Y' = 1.72X_1 - 19.44X_2 + 154.54X_3 + 593.03X_4 - 2,232.92X_5 - 10,317.12$$

The negative values of b_2 and b_5 indicate that the slopes for these variables are inverse. In other words, previous experience shows a negative relationship with current salary. This means that individuals with fewer months of previous experience tend to receive higher current salaries. This may reflect salary inflation with people who were hired more recently receiving higher salaries. The inverse slope associated with the coded gender variable indicates that women are paid lower current salaries (even after all of the other predictors in the model are statistically controlled).

Using the Regression Equation to Make Predictions

We can now use the regression equation to predict people's current salaries on the basis of their beginning salaries, months of previous experience, months since hire, years of education, and gender. Let's say that Evan was paid a beginning salary of $25,000. He has 24 months of previous experience, was hired 12 months ago, has 18 years of education, and is a man. What would we predict his current salary to be? To answer this question, we simply need to substitute these values into the corresponding X values in our regression equation and then solve the equation. Since beginning salary is labeled X_1, previous experience X_2, months since hire X_3, educational level X_4, and gender X_5, we need to substitute 25,000 for X_1, 24 for X_2, 12 for X_3, 18 for X_4, and 1 (the numeric code we assigned for men) for X_5 and then solve. Let's go ahead and do that:

$1.7228(25,000) - 19.4361(24) + 154.5362(12) + 593.0313(18) - 2,232.9171(1) - 10,317.1153 = 42,582.50$

Thus, we would predict Evan's current salary to be $42,582.50.

Reporting the Results

We could now report the following results:

> A stepwise regression analysis was conducted using beginning salary, years of education, months since hire, previous experience, gender, and minority classification to predict current salary. The results revealed a significant five-predictor model, $R^2 = .81$, $F(5, 468) = 408.69$, $p < .001$.

Specifically, beginning salary, $b = 1.72$, $SE_b = 0.06$, $\beta = .79$, $t(468) = 28.47$, $p < .001$; months of previous experience, $b = -19.44$, $SE_b = 3.58$, $\beta = -.12$, $t(468) = -5.42$, $p < .001$; months since hire, $b = 154.54$, $SE_b = 34.08$, $\beta = .09$, $t(468) = 4.53$, $p < .001$; years of education, $b = 593.03$, $SE_b = 166.63$, $\beta = .10$, $t(468) = 3.56$, $p < .001$; and gender, $b = -2{,}232.92$, $SE_b = 792.08$, $\beta = -.07$, $t(468) = -2.82$, $p = .005$; were all significant predictors of current salary. Minority classification was not included in the model as its inclusion would not have significantly improved on its predictive utility, $\beta = -.03$, $t(467) = -1.62$, $p = .11$.

Try On Your Own

Try to use stepwise regression to predict beginning salary using educational level, previous experience, months since hire, gender, and minority classification. Build the least squares regression equation, and report the results in APA style.

Sign Test

Learning Objectives

In this chapter, you will learn how to analyze data using the sign test. You will learn how to conduct the analysis, interpret, and report the results for both two-tailed and one-tailed tests.

INTRODUCTION TO THE SIGN TEST

The sign test can be used to analyze data from a within-subjects (aka repeated measures) design with two conditions. It can be used to determine whether there is a significant difference in the direction of responses across the two conditions of an experiment. Specifically, the difference in each participant's scores on the dependent variable across the two conditions is expressed as a plus or a minus sign, and the total number of plus and minus signs are compared to determine whether there is a significant difference. Unlike the *t*-test (described in Chapter Eight), it does not carry the assumptions that the data are normally distributed or that the dependent variable was measured on an interval or ratio scale. Since it does not carry these assumptions, it is referred to as a nonparametric test.

TWO-TAILED SIGN TEST

Imagine you have just been hired by campus food services to conduct a Pepsi Challenge in order to determine whether students on your campus prefer the taste of Pepsi or Coke. You predict that one brand will be preferred over the other. Since you do not predict which brand of soft drink will be preferred, your hypothesis is considered nondirectional, and you will need to perform a two-tailed sign test. Before conducting the experiment, you decide to stick with convention and set alpha at .05.

You go to the Student Union Building and ask 12 students to do a blind taste test. You ask each student to take a sip of Pepsi and a sip of Coke and indicate which they prefer. To guard against order effects, you use complete counterbalancing. You randomly select six students to take a sip of Pepsi first and a sip of Coke second and six students to take a sip of Coke first and a sip of Pepsi second. Assume you obtain the results shown below. A "1" indicates that the participant preferred the soft drink, and a "2" means they didn't prefer the soft drink. You should be able to see that the participant with the ID code 1 preferred Pepsi, while the participant with ID code 4 preferred Coke.

ID	Coke	Pepsi	ID	Coke	Pepsi
1	2	1	7	2	1
2	2	1	8	1	2
3	2	1	9	2	1
4	1	2	10	2	1
5	2	1	11	2	1
6	2	1	12	2	1

Begin by **opening a blank SPSS spreadsheet** and **entering** the **data** into three columns and 12 rows in the **Data View window** (refer to the "Entering Data" section in Chapter One if you forget how to do this). Note that the top row contains the **variable names**, which must be entered in the **Variable View window**. While you are in Variable View, you should label the values of the Coke and Pepsi variables using the **Values** column. Label **1** as **Preferred** and **2** as **Not Preferred** and define their scale of **Measure** as **Ordinal** (refer to the "Creating Variables" section in Chapter One if you forget how to do this). Once the data are entered, your Data View window should look like the one shown here.

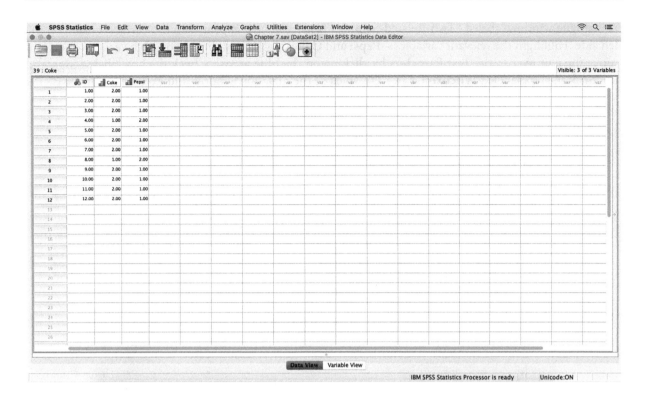

Conducting a Sign Test (Analyze→Nonparametric Tests→Legacy Dialogs→2 Related Samples)

To initiate the sign test analysis, use the upper toolbar to go to **Analyze→Nonparametric Tests→Legacy Dialogs→2 Related Samples**.

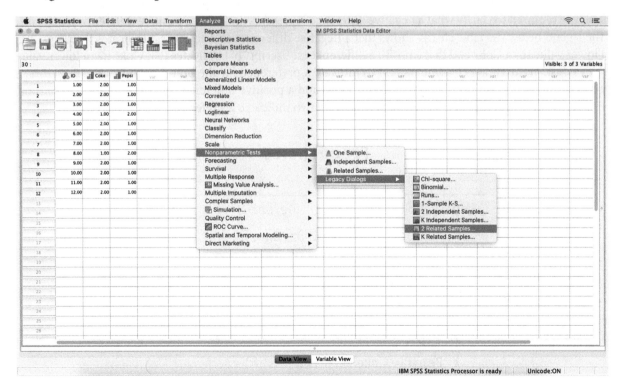

A dialogue window labeled "Two-Related-Samples Tests" will now appear with the variables listed on the left side. Highlight the relevant variables—**Pepsi** and **Coke**—by simply clicking on each variable once, and move the pair over to the **Test Pairs box** by clicking on the **blue arrow**. **Check** the box labeled **Sign** and **uncheck** the box labeled **Wilcoxon**. Click **OK** to close the dialogue window and execute the analysis.

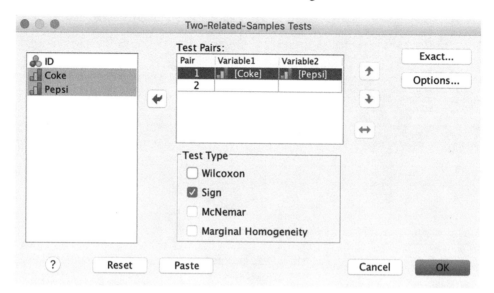

Interpreting the Results

An output window will now appear, displaying two tables. We will begin by examining the "Frequencies" table shown below. This table displays the total number of negative differences (minus signs), the total number of positive differences (plus signs), and the total number of ties. Since the signs are meaningless unless you know which variable was subtracted from which, the subscripts "a," "b," and "c" presented in the table are required to clarify what a negative difference represents and what a positive difference represents. The legend below the table clearly shows that a negative difference (subscript "a") indicates that Pepsi has a lower value than Coke and that a positive difference (subscript "b") indicates that Coke has a lower value than Pepsi. We used lower numbers (1s) to indicate a preference for the brand of soda, and, therefore, a negative difference indicates a preference for Pepsi and a positive difference a preference for Coke. Thus, 10 people indicated a preference for Pepsi, and only two indicated a preference for Coke.

Frequencies

		N
Pepsi – Coke	Negative Differences[a]	10
	Positive Differences[b]	2
	Ties[c]	0
	Total	12

a. Pepsi < Coke

b. Pepsi > Coke

c. Pepsi = Coke

The output window also contains the "Test Statistics" table shown below. This table displays the obtained p value for a two-tailed test, which is the appropriate test to use because our hypothesis was nondirectional (we did not predict which brand would be preferred). The p value represents the probability we would obtain these results or results more extreme on the basis of chance alone (the probability we would obtain 10 or more minuses from 12 participants if the two brands of soft drinks are actually equally preferred in the population). The table shows that the p value is .04. Thus, the probability that this particular result (10 negative differences) or results more extreme would be obtained on the basis of chance alone is .04. Since $p < .05$, we conclude that the two brands of soft drinks are not equally preferred (i.e., that one brand is preferred over the other). Moreover, since the number of negative differences outweighs the number of positive differences, and negative differences represent a preference for Pepsi, we can further conclude that Pepsi is preferred over Coke, in taste tests.[1]

Test Statistics[a]

	Pepsi – Coke
Exact Sig. (2-tailed)	.039[b]

a. Sign Test

b. Binomial distribution used.

Reporting the Results

We could report these results simply by stating the following:

A sign test revealed a significant difference in soft drink preference, with 10 of the 12 participants preferring Pepsi to Coke ($p = .04$).

ONE-TAILED SIGN TEST

In the previous example, we used a nondirectional hypothesis and two-tailed test. For this next example, we will use a directional hypothesis (we will predict the direction of effect) and a one-tailed test.

Imagine you have been hired by a drug company to test the effectiveness of a new sleeping medication called zzz-Aid. Specifically, the company wants to know whether zzz-Aid is more effective (i.e., puts people to sleep faster) than a placebo. Before conducting the study, you decide to set alpha at .05 (one-tailed). You randomly sample 18 individuals from the community and invite each to your sleep lab for two nights. You give each individual zzz-Aid on one night and a placebo on the other night. To guard against order effects, you use complete counterbalancing. You randomly select nine participants to take zzz-Aid on the first night and a placebo on the second night, and nine participants to take the placebo on the first night

---

[1] While the data contained in this guide are fabricated, some interesting research on the Pepsi challenge was conducted because Pepsi typically comes out on top in taste tests, but Coke tends to sell more product than Pepsi. Pepsi is sweeter than Coke, and the results of research indicate that people prefer sweeter drinks when taking only a sip, but when drinking an entire can they prefer Coke, which is less sweet. Unfortunately, this discovery was made after Coke spent millions on the sweeter tasting "New Coke" in the 1980s—a huge failure for them! For those of you who have taken a research methods course, you may recognize this as an issue of external validity. Taste tests have problems with external validity because in real life people tend to drink more than a sip.

and zzz-Aid on the second night. Participants are not told which night they are given the placebo or the drug. You measure and record the number of minutes it takes each participant to fall asleep on each night. Assume you obtained the data shown below.

ID	zzzAid	Placebo	ID	zzzAid	Placebo
1	12	21	10	16	19
2	9	16	11	10	15
3	11	8	12	15	9
4	21	36	13	10	22
5	17	28	14	28	32
6	22	20	15	8	12
7	18	29	16	20	18
8	11	22	17	12	12
9	10	21	18	8	8

Begin by **opening a blank SPSS spreadsheet** and **entering the data** into three columns and 18 rows in the Data View window. Once again, the top row contains the **variable names**, which must be entered in the **Variable View window**. After the data are entered, your Data View window should look like the one displayed here.

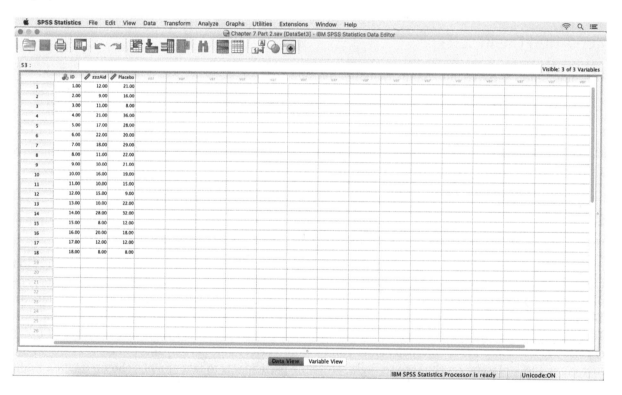

Conducting a Sign Test (Analyze→Nonparametric Tests→Legacy Dialogs→ 2 Related Samples)

The method for conducting a one-tailed sign test is the same as that for conducting a two-tailed sign test. Using the upper toolbar go to **Analyze→Nonparametric Tests→Legacy Dialogs→2 Related Samples.**

Using the "Two-Related-Samples Test" dialogue window shown below, highlight the relevant variables—**zzzAid** and **Placebo**—by simply clicking on each variable. Move the pair over to the **Test Pairs box** by clicking on the **blue arrow**. **Check** the box labeled **Sign** and **uncheck** the box labeled **Wilcoxon**. Click **OK**.

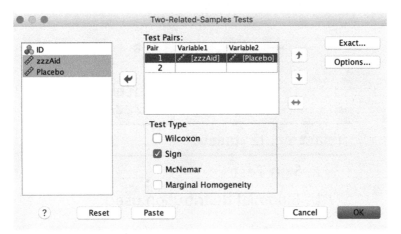

Interpreting the Results

We will once again start by examining the "Frequencies" table that appears in the output window. From this table, you should be able to determine that there are four negative differences, 12 positive differences, and two ties. The legend below the table shows that a negative difference represents lower scores in the placebo condition than in the zzz-Aid condition. Since the data represents the number of minutes it took for participants to fall asleep, this means that four participants fell asleep faster when they took the placebo. The legend below the table also shows that a positive difference represents lower scores in the zzz-Aid condition. Therefore, we can determine that 12 participants fell asleep faster when they took zzz-Aid. The two ties indicate that for two participants there was no difference in the number of minutes it took them to fall asleep. The sign test cannot handle ties, so it simply discards the data from ties. Thus, our sample size (n) is now reduced to 16 participants.

Frequencies

		N
Placebo – zzzAid	Negative Differences[a]	4
	Positive Differences[b]	12
	Ties[c]	2
	Total	18

a. Placebo < zzzAid

b. Placebo > zzzAid

c. Placebo = zzzAid

The "Test Statistics" table displays the p value for a two-tailed test in the row labeled "Exact Sig. (2-tailed)." Recall from Chapter Three that two-tailed tests are used when the hypothesis is nondirectional. However, for this example, we have a directional hypothesis and, therefore, we should use a one-tailed test. SPSS does not provide an option to run a one-tailed sign test; but it is a simple process to manually convert the

p value. To obtain the p value for a one-tailed test, simply divide the p value for the two-tailed test in half. For added precision, you should use a p value that has been rounded to at least four decimal places. By clicking repeatedly on the value in the table, you will find that the p value rounded to four decimal places is .0768. Since .0768 ÷ 2 = .0384, the p value for our one-tailed test is .04.

Test Statistics[a]

	Placebo – zzzAid
Exact Sig. (2–tailed)	.077[b]

a. Sign Test

b. Binomial distribution used.

Remember when we use a directional hypothesis, we can reject the null hypothesis only if the results turn out in the predicted direction. We predicted that zzz-Aid would decrease the number of minutes it takes participants to fall asleep. As described earlier, this is represented as a positive difference. Since the number of positive differences outweigh the number of negative differences, we know the results worked out in the predicted direction. Moreover, since $p < .05$ (one-tailed), we can conclude that the difference is statistically significant.

Reporting the Results

We can report these results simply by stating the following:

> A one-tailed sign test revealed that significantly more participants fell asleep faster after taking zzz-Aid relative to a placebo ($p = .04$, one-tailed).

Predicting the Wrong Direction

You should note that if the results had worked out in the opposite direction to the one we predicted and there were instead 12 negative differences and four positive differences (meaning that for most people zzz-Aid increased the number of minutes it takes to fall asleep), then the displayed p value would be the same, but we would have to make an adjustment to it by subtracting it from 1. In this case, we would report the p value as .96 (1 − .0384 = .96), and since this value is higher than alpha (.05), we would have to conclude that the result is not statistically significant. In this case, we would need to report that the majority of participants did not fall asleep faster after taking zzz-Aid relative to a placebo ($p = .96$, one-tailed). This once again demonstrates the risk of using a directional hypothesis and the importance of examining the direction of effect before reaching a conclusion about the significance of the results of a one-tailed test.

t-TESTS

Learning Objectives

In this chapter, you will learn how to analyze data using Student's single-sample *t*-test, paired-samples *t*-test, and independent-samples *t*-test. For each, you will learn how to conduct the analysis and interpret and report the results for both nondirectional and directional hypotheses. You will also learn to interpret the confidence intervals and the effect-size indicators, Cohen's *d* and Hedges' *g*, for all three types of tests.

SINGLE-SAMPLE *t*-TEST

The single-sample *t*-test, alternatively referred to as the one-sample *t*-test, is used to determine whether the mean of a sample (*M*) is significantly different from the mean of a population (μ). As such, it can be used to infer whether a sample belongs to a population or is unique from a population.

According to Statistics Canada's census results, the mean weight of Canadian women was 155 lb. in 2009. Imagine you are a health researcher interested in examining whether there has been a significant change in the mean weight of Canadian women since 2009. Although obesity rates are currently at an all-time high, you also recognize that many Canadians are beginning to adopt healthier lifestyles. As such, you decide to use a nondirectional hypothesis and simply predict that there has been a significant change in the mean weight of Canadian women since 2009. Before collecting any data, you decide to stick with the convention and set alpha at .05. Assume you randomly sample 30 women from the general population of Canada and obtain the following data on their weights (in lb).

ID	Weight	ID	Weight	ID	Weight
1	165	11	149	21	129
2	180	12	136	22	117
3	107	13	212	23	152
4	163	14	157	24	183
5	195	15	132	25	169
6	159	16	128	26	145
7	168	17	178	27	265
8	145	18	210	28	162
9	125	19	159	29	179
10	198	20	195	30	285

Begin by **opening** a **blank SPSS spreadsheet**. **Enter** the 30 women's **ID codes** and **weights** into two columns in the **Data View window**. Note that the top row contains the **variable names**, which must be entered in the **Variable View window**. While you are in the Variable View window, you should define the properties of the variables. **Label ID** as **ID Codes** and **Weight** as **Weight in lbs.** and define the scales of measure of **ID** as **Nominal** and **Weight** as **Scale** (refer to the "Creating Variables" section in Chapter 1 if you forget how to do this). Once the data are entered, your Data View window should look like the one shown here.

Conducting a Single-Sample *t*-Test (Analyze→Compare Means→One-Sample t Test)

Since we want to compare a sample mean to a population mean to determine whether there is a significant difference, we will need to conduct a single-sample *t*-test. To conduct the analysis, use the upper toolbar to go to **Analyze→Compare Means→One-Sample T Test**.

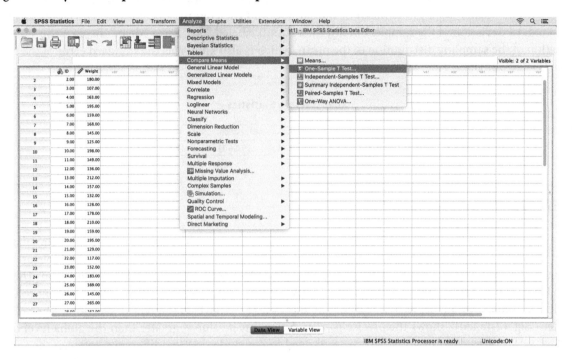

A "One-Sample T Test" dialogue window, like the one shown here, will now appear. Highlight the relevant variable—**Weight**—by clicking on the variable label and move it over to the **Test Variable(s): box** by clicking the **blue arrow**. Next enter the population mean—**155**—into the **Test Value box**. Make sure that **Estimate effect sizes** are **checked**. Click **OK** to close the dialogue window and execute the analysis.

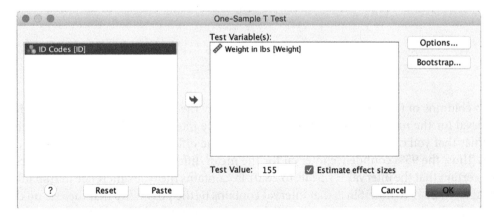

Interpreting the Results

Descriptive Statistics

An output window will now appear, displaying two tables. We will begin by examining the "One-Sample Statistics" table shown below. The first column simply displays the label for the variable you are considering in the analysis. The column labeled "*N*" displays the number of participants in the sample. The column labeled "Mean" displays the mean of the sample, and the column labeled "Std. Deviation" displays the standard deviation of the sample. Finally, the column labeled "Std. Error Mean" displays the standard deviation of the sampling distribution of the mean, which is also known as the standard error of the mean (abbreviated *SEM*).

One-Sample Statistics

	N	Mean	Std. Deviation	Std. Error Mean
Weight in lbs	30	168.2333	39.56024	7.22268

Assessing Statistical Significance

Next, we will consider the "One-Sample Test" table shown below. The primary results of the analysis are presented in this table. The mean of the population is provided at the very top of the table as "Test Value = 155." You can use this to confirm that you correctly entered the mean of the population. The column labeled "t" provides the obtained *t* value. From this we can determine that $t = 1.83$. The next column, labeled "df" displays the degrees of freedom; it shows that we have 29 degrees of freedom. The column labeled "Sig. (2-tailed)" shows the precise *p* value. Since this *p* value is greater than the alpha level of .05 that we set prior to collecting the data, we will need to conclude that the difference between the sample mean and the population mean is not statistically significant. The column labeled "Mean Difference" displays the difference between the sample mean and the population mean. This value indicates that our sample has a mean weight that is 13.23 lb. higher than the mean weight of the population. These results would be reported in the following way: $t(29) = 1.83, p = .08$.

One-Sample Test

Test Value = 155

	t	df	Sig. (2–tailed)	Mean Difference	95% Confidence Interval of the Difference	
					Lower	Upper
Weight in lbs	1.832	29	.077	13.23333	–1.5387	28.0054

The last two columns of the "One-Sample Test" table display the lower and upper limits of the 95% confidence interval for the *mean difference*. The 95% confidence interval for the mean difference provides a range of values that you can be 95% certain contains the true difference between the sample and population means. Thus, the 95% confidence interval for the mean difference displayed below indicates that we can be 95% certain that the interval −1.54 lb. to 28.01 lb. contains the true difference in Canadian women's mean weight from 2009 to now. Since that interval contains 0 (the possibility that there is no difference in the mean weights), we must conclude that the difference is not statistically significant.

Calculating Confidence Intervals for the Population Mean

Researchers are often more interested in obtaining the confidence interval for the *population mean* (rather than the confidence interval for the mean difference). The 95% confidence interval for the population mean provides a range of values that you can be 95% certain contains the true population mean. This is a subtle but important distinction. Although SPSS provides the confidence interval only for the mean difference, the confidence interval for the population mean can easily be hand-calculated. To calculate the 95% confidence interval for the population mean, we will need the sample mean (M), the standard error of the mean (SEM), and the critical t value for an alpha of .05 and 29 degrees of freedom (t_{crit}). As described earlier, the sample mean and the standard error of the mean can both be found in the "One-Sample Statistics" table. To determine the critical t value, you will need to refer to a table of the Critical Values of Student's t distribution. For a two-tailed test with alpha of .05 and 29 degrees of freedom, the critical t value is 2.045. The formulas for calculating the lower and upper limits of the confidence interval for the population mean are provided below. Let's go ahead and practice hand-calculating the confidence interval for the population mean.

$$\mu_{lower} = M - SEM\ (t_{crit}) = 168.2333 - 7.2227(2.045) = 153.46$$

$$\mu_{upper} = M + SEM\ (t_{crit}) = 168.2333 + 7.2227(2.045) = 183.00$$

Since 155—the mean weight of the population of Canadian women in 2009—is contained within this interval, we will once again need to conclude that there has not been a significant change in the mean weight of Canadian women since 2009. Note that using either the traditional (p value) method or the alternative confidence interval method, we will always arrive at the same conclusion about the significance of our findings.

Effect Sizes (Cohen's d and Hedges' g)

Estimates of the effect sizes for t-tests were added to the most recent version of SPSS (Version 27). So, if you are using a recent version of SPSS, then the output will also contain a "One-Sample Effect Sizes" table like the one shown here.

One-Sample Effect Sizes

		Standardizer[a]	Point Estimate	95% Confidence Interval Lower	95% Confidence Interval Upper
Weight in lbs	Cohen's d	39.56024	.335	−.036	.700
	Hedges' correction	40.62148	.326	−.035	.681

a. The denominator used in estimating the effect sizes.
Cohen's d uses the sample standard deviation.
Hedges' correction uses the sample standard deviation, plus a correction factor.

The table provides the Cohen's d value in the row labeled "Cohen's d" and the column labeled "Point Estimate." As shown in the table, the Cohen's d value is 0.33. Cohen's d values of 0.20 are generally considered small, values of 0.50 are generally interpreted as medium, whereas values of 0.80 and higher are typically considered large. More specifically, the Cohen's d value is interpreted as the number of standard deviation units the sample mean is from the population mean. As such, our Cohen's d value of 0.33 indicates that the effect is small. More specifically, this value indicates that our sample mean is only 0.33 standard deviation units above our population mean. Note that because Cohen's d values can be greater than 1, a leading zero should be used before the decimal place when reporting values below 1, and the d should be italicized.

Often, the numerator of the formula for Cohen's d contains the *absolute* difference in the means so that Cohen's d is always a positive value. SPSS uses a formula with the sample mean minus the population mean in the formula, rather than considering the absolute difference in these values. As such, the displayed Cohen's d value in the SPSS output may be positive or negative. It will be positive when the sample mean is higher than the population mean, and it will be negative when the population mean is higher than the sample mean. You can just remove the sign from the value in cases where it is displayed as negative and report its absolute value. The column labeled "Standardizer" simply shows the value in the denominator of the formula used by SPSS to compute the effect-size indicator. In the case of Cohen's *d*, this value is the sample standard deviation, which you can confirm for yourself by referring to the standard deviation reported in the "One-Sample Statistics" table previously displayed.

The table also displays a point estimate and confidence interval for Hedges' *g* in the row labeled "Hedges' correction." Hedges' *g* is another commonly used indicator of effect sizes for *t*-tests that is closely related to Cohen's *d*. As you can see in the column labeled "Standardizer," Hedges' *g* contains a somewhat larger value in the denominator than Cohen's *d*, and as such, Hedges' *g* will be somewhat smaller than Cohen's *d*. Hedges' *g* is considered a less biased effect-size indicator and is recommended to use when the sample size is less than 20. With larger samples, the difference between Cohen's *d* and Hedges' *g* will typically be trivial. Indeed, in our case the two values round to the same value (0.33). Hedges' *g* is interpreted in a similar manner as Cohen's *d* with values of 0.20 considered small, values of 0.50 interpreted as medium, and values of 0.80 and higher considered large-sized effects.

Finally, the "One-Sample Effect Sizes" table also contains the lower and upper limits of the 95% confidence interval for Cohen's *d* and Hedges' *g*. As you can see, the confidence intervals for this example ranges from -0.04 to 0.70 for Cohen's *d* and from -0.04 to 0.68 for Hedges' *g*. The fact that these confidence intervals cross 0 is further indication that the difference in the sample and population means is not statistically significant (it contains the possibility that the effect size is 0, indicating no effect).

Reporting the Results

On the basis of these results, we could report the following:

> A single-sample *t*-test revealed that the mean weight of today's sample of Canadian women ($M = 168.23$, $SD = 39.56$) is not significantly different than the mean weight of the population of Canadian women in 2009 ($\mu = 155$), $t(29) = 1.83$, $p = .08$, $d = 0.33$, 95% CI [153.46, 183.00].

ONE-TAILED SINGLE-SAMPLE *t*-TEST

Let's now assume that you are a dietician who has growing concerns about the impact of the rapid growth of the fast food industry on Canadian women's weight. As such, you are only interested in whether the mean weight of Canadian women has *increased* since 2009. Because your hypothesis is now directional, a one-tailed test is appropriate to use. Assume that prior to collecting any data you decide to set alpha at .05 (one-tailed).[1] You randomly sample the same 30 Canadian women whose data appear at the beginning of this chapter.

SPSS does not provide an option to conduct one-tailed *t*-tests, so directional hypotheses do not influence the way we conduct the analysis. The only thing that differs with directional hypotheses is the manner in which we report the *p* value displayed in the "One-Sample Test" table. The *p* value provided in the table is for a two-tailed test (nondirectional hypothesis). To adjust it for our one-tailed test (directional hypothesis), all we need to do is divide it in half. Let's go ahead and do that:

$$p = .0772 \div 2 = .0386$$

---

[1] Note that this directional hypothesis would have to have been made before collecting any data. It is scientific cheating to switch from a nondirectional to a directional hypothesis after seeing the results! In the long run, practicing this type of cheating will inflate your Type I error rate from 5% to 7.5%.

Thus, our p value for a one-tailed test is .04. Before we reach any conclusions, we need to examine whether our results are in the predicted direction. Our obtained sample mean (shown in the "One-Sample Statistics" table) is 168.23, and our population mean (shown at the very top of the "One-Sample Test" table) is 155. Because our sample mean is higher than our population mean, we know that the results are in the predicted direction.

Reporting the Results

Based on these findings, we could report the following:

> A one-tailed single-sample t-test revealed that the mean weight of the sample of Canadian women ($M = 168.23$, $SD = 39.56$) is significantly higher than the mean weight of the population of Canadian women in 2009 ($\mu = 155$), $t(29) = 1.83$, $p = .04$ (one-tailed), $d = 0.33$.

Predicting the Wrong Direction

It is important to note what would change if we had made a prediction in the wrong direction. Let's assume we had predicted a *decrease* in the mean weight of Canadian women and obtained the same results we've been considering here (showing an *increase* in the mean weight of Canadian women). In this case, our calculation of the p value would change. Rather than simply dividing the p value by 2, as we did when the results worked out in the predicted direction, we would now need to subtract the divided p value from a value of 1. So, we would still divide the p value in half ($0.0772 \div 2 = 0.0386$), but now we would also need to subtract the result from a value of 1 ($1 - 0.0386 = 0.96$). In this case, we would report the following t-test results, $t(29) = 1.83$, $p = .96$, $d = 0.33$. Since $p > .05$ in this scenario, we would need to conclude that there has not been a significant decrease in the mean weight of Canadian women since 2009.

PAIRED-SAMPLES t-TEST

The paired-samples t-test (alternatively referred to as the correlated groups t-test) is used to analyze data from within-subjects (repeated measures) designs with two conditions. Specifically, it is used to determine whether the means of two conditions, containing the same participants, differ significantly.

Assume you are a researcher interested in the acute effects of tetrahydrocannabinol (THC, the primary psychoactive constituent in cannabis) on memory test performance. You use a nondirectional hypothesis and simply predict that THC will affect memory test performance. Before conducting the study, you decide to set alpha at .05. Assume that you randomly select and invite 20 individuals to your lab on two separate days. Each individual is given a joint containing THC to smoke on one day and a placebo joint that doesn't contain any THC to smoke on the other day. To guard against order effects, you use complete counterbalancing. Half of the participants smoke the joint containing THC on Day 1 and the placebo joint on Day 2, and the other half of the participants smoke the placebo joint on Day 1 and the joint containing THC on Day 2. The participants are not informed which day they smoke the joint that contains the THC. After smoking each of the joints, the participants are given a memory test. They are read a list of 15 words and are asked to recall as many words from it as possible. You measure the number of words each participant correctly recalls.

Assume you obtain the data shown below. The numbers in the THC and Placebo columns refer to the number of words correctly recalled after smoking the respective type of joint.

ID	THC	Placebo	ID	THC	Placebo
1	7	8	11	6	8
2	5	7	12	5	7
3	4	5	13	9	12
4	12	10	14	12	12
5	8	9	15	15	12
6	9	11	16	7	9
7	11	13	17	5	7
8	6	5	18	9	11
9	7	9	19	10	8
10	8	8	20	5	6

Begin by **entering** the **data** into a **blank SPSS spreadsheet**. Note that the top row contains the variable names, which must be entered in the Variable View window. The remaining data must be entered into three columns and 20 rows in the Data View window. Keep in mind that different variables must be entered into separate columns, whereas the data for each of the participants must be entered in the rows (each row must therefore represent one participant's data on all of the variables). Once the data have been entered, your Data View window should look like the one displayed here.

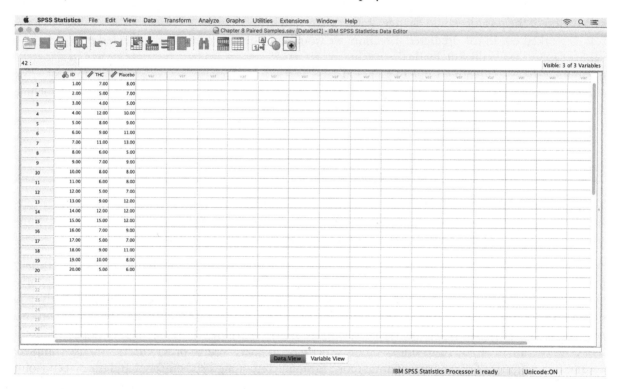

Conducting a Paired-Samples *t*-Test (Analyze→Compare Means→Paired-Samples T Test)

To conduct the paired-samples *t*-test, go to **Analyze→Compare Means→Paired-Samples T Test.**

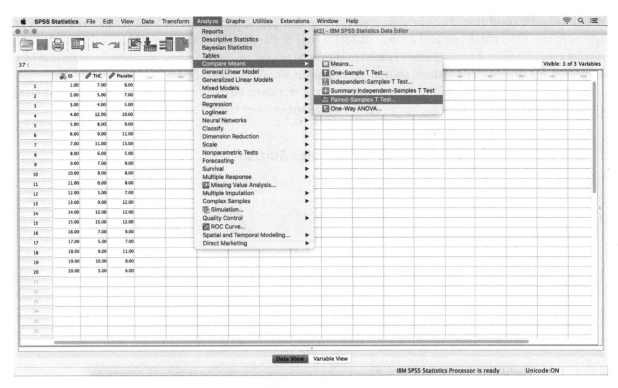

A "Paired-Samples T Test" dialogue window, like the one shown here, will now appear. Highlight the relevant variables—**THC** and **Placebo**—by clicking on each of the variable names, and then click on the **blue arrow** to move them over to the **Paired Variables: box**. Click **OK** to close the "Paired-Samples T Test" dialogue window and initiate the analysis. Make sure that **Estimate effect sizes** is **checked** and that the option to calculate the standardizer using the **Standard deviation of the difference** is **selected.**

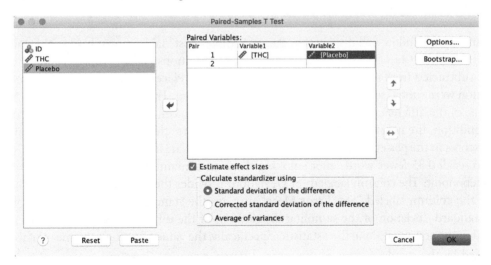

Interpreting the Results

Descriptive Statistics

A series of tables will now appear in the output window. We will first examine the "Paired Samples Statistics" table shown here. This table provides some basic descriptive statistics. The first column in the table displays the names of the conditions that are being compared. The column labeled "Mean" presents the mean for each condition. You should be able to see that participants were able to recall a mean of 8.00 words in the THC condition and a mean of 8.85 words in the placebo condition. The column labeled "N" displays the number of participants in each condition. Since we are considering data from a within-subjects (repeated measures) design, there should always be an equal number of participants in each of the conditions. The column labeled "Std. Deviation" lists the standard deviation for each condition. Finally, the column labeled "Std. Error Mean" provides the standard error of the mean for each condition.

Paired Samples Statistics

		Mean	N	Std. Deviation	Std. Error Mean
Pair 1	THC	8.0000	20	2.90191	.64889
	Placebo	8.8500	20	2.39022	.53447

The next table we will examine is the "Paired Samples Correlation" table shown below. This table displays the correlation between the scores in the two conditions. You learned all about correlation in Chapter 3, so you should be able to see by looking at this table that there is a large positive correlation, $r(18) = .82$, $p < .001$, between the number of words participants were able to recall after smoking a joint containing THC and the number of words they were able to recall after smoking a placebo joint.

Paired Samples Correlations

		N	Correlation	Sig.
Pair 1	THC & Placebo	20	.819	.000

Assessing Statistical Significance

Finally, your output window will display the "Paired Samples Test" table shown below. This table provides the main results of the t-test. The first column displays the conditions being compared and indicates which scores were subtracted from which. The column shows "THC - Placebo," indicating that scores in the placebo condition were subtracted from scores in the THC condition. The next column, labeled "Mean," presents the mean of the difference scores. Since scores in the placebo condition were subtracted from scores in the THC condition, the negative difference score ($-.85$) indicates that scores in the THC condition were lower than scores in the placebo condition. More precisely, the value indicates that, on average, participants were able to recall 0.85 fewer words after smoking the joint containing THC compared with after smoking the placebo joint. The column labeled "St Deviation" provides the standard deviation of the difference scores, and the column labeled "Std. Error Mean" displays the standard error of the mean, which is the estimated standard deviation of the sampling distribution of the difference between sample means. This value is used in the computation of the t statistic. Specifically, the value of t equals the mean of the difference scores divided by the standard error of the mean difference scores. Therefore, $t = \frac{-0.8500}{0.3719} = 2.29$. The column labeled "t" displays the obtained t value. The column labeled "df" gives the degrees of freedom, and the column labeled "Sig. (2-tailed)" provides the p value for a two-tailed test. Since $p < .05$, we can once again conclude that THC has a significant acute effect on memory test performance, $t(19) = -2.29$, $p = .03$.

Paired Samples Test

		Paired Differences							
					95% Confidence Interval of the Difference				
		Mean	Std. Deviation	Std. Error Mean	Lower	Upper	t	df	Sig. (2-tailed)
Pair 1	THC – Placebo	–.85000	1.66307	.37187	–1.62834	–.07166	–2.286	19	.034

The section of the "Paired Samples Test" table labeled "95% Confidence Interval of the Difference" provides the lower and upper limits of the 95% confidence interval for the difference in means. As shown in the table, this interval ranges from -1.63 to -0.07. This indicates that we can be 95% certain that the interval -1.63 to -0.07 contains the true acute effect of THC on memory test performance. In other words, we can be 95% certain that THC decreases memory test performance by 0.07 to 1.63 recalled words. Since the confidence interval does not cross 0, we can conclude that THC has a significant acute effect on memory test performance.

Effect Sizes (Cohen's d and Hedges' g)

Finally, the "Paired Samples Effect Sizes" table shown here contains the point estimates and confidence intervals for Cohen's d and Hedges' g. As displayed in the table in the row labeled "Cohen's d" and the column labeled "Point Estimate," the Cohen's d estimate is 0.51. While a negative value is displayed in the table, typically the absolute value is reported. The Cohen's d value of 0.51 indicates that there is a medium-sized effect of THC on memory test performance. More specifically, it indicates that the mean of the THC condition is 0.51 standard deviations units lower than the mean of the placebo condition. The confidence interval for Cohen's d, reported in the section of the table labeled "95% Confidence Interval," is -0.97 to -0.04. The fact that this interval does not cross 0 further confirms that this effect is statistically significant (it does not contain the possibility that there is no effect of THC on memory test performance).

The value of Hedges' g and the 95% confidence interval for Hedges' g are reported in the row labeled "Hedges' correction." Again, Hedges' g provides a less biased indicator of the effect size and is typically slightly smaller than the Cohen's d value. Again, it is recommended to report Hedges' g when the sample size is less than 20.

Paired Samples Effect Sizes

| | | | Standardizer[a] | Point Estimate | 95% Confidence Interval | |
					Lower	Upper
Pair 1	THC – Placebo	Cohen's d	1.66307	-.511	-.972	-.038
		Hedges' correction	1.69682	-.501	-.953	-.037

a. The denominator used in estimating the effect sizes.
Cohen's d uses the sample standard deviation of the mean difference.
Hedges' correction uses the sample standard deviation of the mean difference, plus a correction factor.

Reporting the Results

Bringing it all together can now report the following:

> A paired-samples t-test revealed that the mean number of words participants could recall after smoking the joint containing THC ($M = 8.00$, $SD = 2.90$) was significantly lower than the mean number of words they could recall after smoking the placebo joint ($M = 8.85$, $SD = 2.39$), $t(19) = -2.29$, $p = .03$, $d = 0.51$, 95% CI $[-1.63, -0.07]$.

Calculating the 99% Confidence Intervals for the Difference in Means

By default, SPSS provides the 95% confidence interval because it is the most commonly used confidence interval. However, you can change this setting and request any confidence interval you'd like. We'll practice by changing the setting to give us the second most frequently used confidence interval, the 99% confidence interval. This interval corresponds to an alpha of .01.

To change the confidence interval, go to **Analyze→Compare Means→Paired-Samples T Test.** The variables—**THC** and **Placebo**—should still appear in the **Paired Variables box.** (If they do not, then place them in that box using the blue arrow.) Next, click the **Options** box. A "Paired-Samples T Test: Options" box like the one shown here will now appear. Replace the value of 95 with a value of **99** by simply typing in the box. Press **Continue** to close that dialogue window. Finally, click **OK** to close the "Paired-Samples T Test" dialogue window and initiate the analysis.

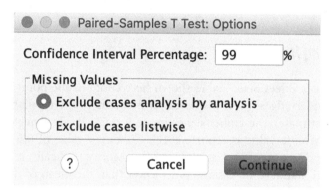

The same four tables will appear in the output window. The only difference between these tables and the ones produced previously are the confidence intervals. You should note that nothing else in the tables have changed. The "Paired Samples Test" table now displays the 99% confidence interval for the difference in means. According to this confidence interval, we can be 99% certain that the interval -1.91 to 0.21 contains the true acute effect of THC on memory test performance. Similarly, the "Paired Samples Effect Sizes" table will display the 99% confidence intervals for Cohen's d and Hedges' g. Since all of these confidence intervals contain the possibility that THC has no effect on memory test performance (the intervals all cross 0), we must conclude that THC does not have a significant effect on memory test performance. Note that we would have reached this same conclusion using the traditional p value method and an alpha of .01 (the alpha level that corresponds to the 99% confidence interval) because $p = .03 > .01$.

Paired Samples Test

| | | | Paired Differences | | | | | | |
| | | Mean | Std. Deviation | Std. Error Mean | 99% Confidence Interval of the Difference | | t | df | Sig. (2–tailed) |
					Lower	Upper			
Pair 1	THC – Placebo	-.85000	1.66307	.37187	-1.91390	.21390	-2.286	19	.034

Reporting the Results

The results using an alpha of .01 and the 99% confidence interval could be reported in the following manner:

> Using an alpha of .01, a paired-samples t-test revealed that the mean number of words participants could recall after smoking the joint containing THC ($M = 8.00$, $SD = 2.90$) was not significantly different than the mean number of words they could recall after smoking the placebo joint ($M = 8.85$, $SD = 2.39$), $t(19) = -2.29$, $p = .03$, $d = 0.51$, 99% CI $[-1.91, 0.21]$.

ONE-TAILED PAIRED-SAMPLES t-TEST

Now assume you had originally made a directional hypothesis and predicted that THC would *decrease* memory test performance. In this scenario, you would need to conduct a one-tailed paired-samples t-test.

The method of conducting the analysis would remain the same, and the tables in the output would remain the same. The only thing that would change in this scenario is the reported p value, which would once again be cut in half. The p value for the two-tailed test was .0339, so the p value for a one-tailed test would be .02 (.0339 $\div$ 2 = .0170). Since the mean of the THC condition was lower than the mean of the placebo condition, we know the results are in the predicted direction.

Reporting the Results

Returning to the scenario of a one-tailed paired-samples t-test (with alpha of .05), we would report the following:

> A one-tailed paired-samples t-test revealed that the mean number of words participants could recall after smoking the joint containing THC ($M = 8.00$, $SD = 2.90$) was significantly lower than the mean number of words they could recall after smoking the placebo joint ($M = 8.85$, $SD = 2.39$), $t(19) = -2.29$, $p = .02$ (one-tailed), $d = 0.51$.

Predicting the Wrong Direction

Again, if the results had worked out in the opposite direction to the one that we predicted and showed that memory test scores were actually higher in the THC condition relative to the placebo condition, then we would need to further adjust the p value. Rather than simply dividing the p value in half, as we did when the results worked out in the predicted direction, we would now need to subtract the adjusted p value from a value of 1. So, we would still divide the p value in half (.0339 $\div$ 2 = .0170), but now we would also need to subtract the result from a value of 1 (1 $-$ .0170 = .98) because in this scenario the results are opposite to what we predicted. In this case, we would report that THC did not have a significant negative effect on memory test performance, $t(19) = 2.29$, $p = .98$, $d = 0.51$.

INDEPENDENT-SAMPLES t-TEST

The independent-samples t-test is also used to determine whether the means of two conditions differ significantly. In contrast to the paired-samples t-test, it is used in conjunction with between-subjects (independent groups) designs in which the two conditions contain different groups of participants.

A previous president of Harvard University once publicly speculated that one reason why women are underrepresented in the sciences is that there is a "different availability of aptitude at the high end." In other words, he claimed that women are underrepresented in the sciences because they do not have the innate abilities necessary to excel in them. Let's assume you are a researcher who is infuriated by this statement because you think statements like these adversely affect women. As such, you decide to examine the influence of claims like this one on women's actual math test performance.

Assume you invite 20 women to your lab and randomly assign each woman to one of two separate conditions. Women in both conditions are told to wait in a waiting area where they "accidentally overhear" a staged conversation between two professors. Women in condition 1 (the genetics condition) overhear the professors discussing a new paper revealing evidence that a math gene has been discovered that is sex-linked and commonly deficient in women and that this may account for why women are underrepresented in the sciences. Women in condition 2 (the socialization condition), overhear the professors discussing a new paper revealing evidence that women only tend to be underrepresented in the sciences because they are raised in a society where they are regarded as being inferior at math. After participants overhear one of these staged conversations, they are asked to complete a math test containing 20 questions. You hypothesize that women's math test performance will be affected by the message they overhear. Prior to collecting any data, you decide to set alpha at .05. Assume you record the following data. The codes "1" and "2" under the columns labeled "Message" indicate which condition participants were assigned to, and the values in the "MathScore" columns indicate how many math questions each woman got correct.

ID	Message	MathScore	ID	Message	MathScore
1	1	15	11	1	8
2	2	18	12	2	20
3	1	12	13	1	12
4	2	13	14	2	16
5	1	11	15	1	14
6	2	10	16	2	11
7	1	15	17	1	9
8	2	17	18	2	16
9	1	14	19	1	16
10	2	19	20	2	20

Begin by **entering** the **data** into a **blank SPSS spreadsheet**. Note that the top row contains the variable names, which must be entered in the Variable View window. While you are in Variable View, use the **Values** column to identify message 1 as **Genetics** and message 2 as **Socialization**. Define the scales of measure for **ID** and **Message** as **Nominal** and for **MathScore** as **Scale**. The remaining data must be entered into three columns and 20 rows in the Data View window. Keep in mind that different variables must be entered into separate columns, whereas the data for each of the participants must be entered in the rows. Once the data are entered, your Data View window should look like the one displayed here.

Conducting an Independent-Samples *t*-Test (Analyze→Compare Means→Independent-Samples T Test)

To conduct an independent-samples *t*-test, you need to use the upper toolbar to go to **Analyze→Compare Means→Independent-Samples T Test**.

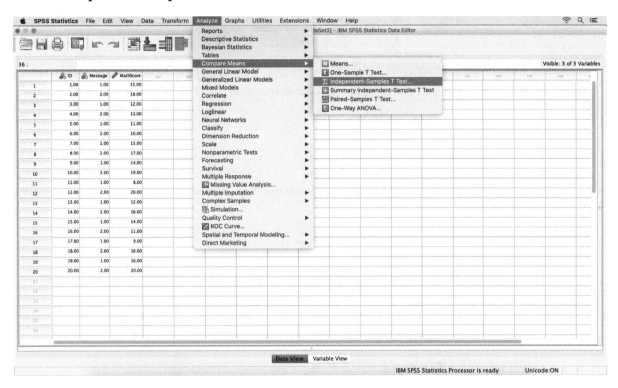

This will open up an "Independent-Samples T Test" dialogue window, like the one shown here. Highlight the dependent variable—**MathScore**—by clicking on the variable name and move it over to the **Test Variable(s): box** using the corresponding **blue arrow**. Next, highlight the independent variable—**Message**—by clicking on the variable name, and move it over to the **Grouping Variable: box** by clicking the corresponding **blue arrow**. Make sure the option to **Estimate effect sizes** is **checked**. Next, click on the **Define Groups tab**.

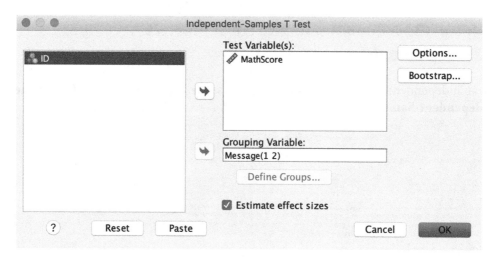

A "Define Groups" dialogue window will now appear. Since we used 1s and 2s to identify the two conditions, you should enter **1** into the box labeled **Group 1** and **2** into the box labeled **Group 2**. Close the dialogue window by clicking **Continue**. Finally, click **OK** to close the "Independent-Samples T Test" dialogue window and initiate the analysis.

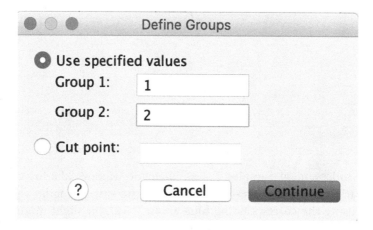

Interpreting the Results

Descriptive Statistics

An output window will now appear, displaying the following tables. We will first examine the "Group Statistics" table. This first column of the table displays the dependent variable under consideration (Math-Score) and the labels we provided for the two conditions. The column labeled "N" shows the number of participants in each condition (there were 10 participants in each condition). The column labeled "Mean" displays the mean math test score in each of the two conditions. You should be able to see that the mean math test score of women in the genetics condition was 12.60, and the mean math test score of women in the socialization condition was 16.00. Note that women's scores were lower in the genetics condition. The column labeled "Std. Deviation" shows the standard deviation of scores in each condition. Finally, the column labeled "Std. Error Mean" provides the standard error of the mean for each condition.

Group Statistics

	Message	N	Mean	Std. Deviation	Std. Error Mean
MathScore	Genetics	10	12.6000	2.67499	.84591
	Socialization	10	16.0000	3.59011	1.13529

Levene's Test of Homogeneity of Variance

By default, SPSS tests the assumption of homogeneity of variance that underlies the independent-samples *t*-test. This is the assumption that the variances of the populations from which the samples were drawn are not significantly different. The section of the "Independent Samples Test" table labeled "Levene's Test for Equality of Variances" contains the results of the test of this assumption. If the results of Levene's test are not significant ($p > .05$), then the variances are not significantly different, meaning the assumption has been met. If the results of Levene's test are significant ($p < .05$), then the variances are significantly different, meaning the data violate the assumption of homogeneity of variance. We do not worry too much about violating this assumption because the independent-samples *t*-test is robust to (i.e., affected very little by) violations of this assumption, especially when an equal number of participants are used in each condition. SPSS suggests you report the adjusted values in the bottom row (the row labeled "Equal variances not assumed") when the assumption has been violated. Note you should also indicate when the assumption was violated in your report. You can see by looking at the column labeled "Sig." that for our analysis the result of Levene's test was not significant ($p = .45$). Hence, we can conclude that the variances are not significantly different and, as such, that the assumption of homogeneity of variance has been met.

Independent Samples Test

		Levene's Test for Equality of Variances		t-test for Equality of Means					95% Confidence Interval of the Difference	
		F	Sig.	t	df	Sig. (2-tailed)	Mean Difference	Std. Error Difference	Lower	Upper
MathScore	Equal variances assumed	.604	.447	-2.401	18	.027	-3.40000	1.41578	-6.37445	-.42555
	Equal variances not assumed			-2.401	16.639	.028	-3.40000	1.41578	-6.39199	-.40801

Assessing Statistical Significance

The primary results of the *t*-test are provided in the "Independent Samples Test" table displayed above. The column labeled "t" displays the obtained *t* value. The column labeled "df" gives the degrees of freedom, and the column labeled "Sig. (2-tailed)" provides the *p* value for a two-tailed test. Since the *p* value provided in the table is less than .05, we can conclude that women's math test performance was significantly affected by the message they overheard.

The next piece of information we can extract from the table is the difference in the means of the two groups, which is presented in the column labeled "Mean Difference." The column labeled "Std. Error Difference" provides the standard error of the mean for the difference scores (which is the estimated standard deviation of the sampling distribution of the difference between sample means). Once again, the value of *t* equals the mean difference divided by the standard error of the mean of the difference scores.

Finally, the last section of the table labeled "95% Confidence Interval of the Difference" displays the lower and upper limits of the 95% confidence interval for the difference in means. The interval reported in the table indicates that we can be 95% certain that the interval -6.37 to -0.43 contains the true effect, that the message that women are genetically predisposed to be poor at math, has on women's actual math test performance. In other words, we can be 95% certain that exposure to genetic explanations for women's math abilities decreases their math performance by 0.43 to 6.37 points. Since the confidence interval does not cross 0, we can conclude that this effect is statistically significant.

Effect Sizes (Cohen's d, Hedges' g, and Glass's Δ)

Finally, the "Independents Samples Effect Sizes" table contains the point estimates and confidence intervals for Cohen's *d*, Hedges' *g*, and Glass's delta (Δ). As displayed in the table in the row labeled "Cohen's *d*" and the column labeled "Point Estimate," the Cohen's *d* estimate is 1.07. Again, although a negative value is displayed in the table, typically the absolute value is reported. This Cohen's *d* of 1.07 indicates that there is a large-sized effect of the message the women overheard on their math test performance. Specifically, it indicates that the women who overheard the bogus message that there is a sex-linked math gene performed 1.07 standard deviations lower on the math test than the women who overheard the message that socialization accounts for women's underrepresentation in the sciences. The confidence interval for Cohen's *d*, reported in the section of the table labeled "95% Confidence Interval," is −2.00 to −0.12. The fact that this interval does not cross 0 further confirms that this effect is statistically significant (it does not contain the possibility that there is no effect of the message on women's math performance).

The value of Hedges' *g* and the 95% confidence interval for Hedges' *g* are reported in the row labeled "Hedges' correction." Again, Hedges' *g* provides a less biased indicator of the effect size and is typically slightly smaller than the Cohen's *d* value. Hedges' *g* is often recommended to use when the sample size is less than 20. Finally, Glass's Δ is reported in the bottom row of the table. As the footnote below the table indicates, this estimate of the effect size uses the control group's standard deviation in the denominator of the formula. As you can see in the column labeled "Standardizer" it used the standard deviation of the socialization condition in the denominator of the formula for Glass's Δ. Glass's Δ is interpreted in the same manner as Cohen's *d* and Hedges' *g* with values of 0.20 considered small, values of 0.50 interpreted as medium, and values of 0.80 and higher considered large-sized effects. Glass's Δ is sometimes recommended when the standard deviations of the two groups are significantly different (i.e., when the results of Levene's test of homogeneity of variance are statistically significant).

Independent Samples Effect Sizes

		Standardizer[a]	Point Estimate	95% Confidence Interval	
				Lower	Upper
MathScore	Cohen's d	3.16579	−1.074	−2.004	−.118
	Hedges' correction	3.30580	−1.028	−1.919	−.113
	Glass's delta	3.59011	−.947	−1.903	.050

a. The denominator used in estimating the effect sizes.
Cohen's d uses the pooled standard deviation.
Hedges' correction uses the pooled standard deviation, plus a correction factor.
Glass's delta uses the sample standard deviation of the control group.

Reporting the Results

We can now report the following:

> An independent-samples *t*-test revealed that the math test scores of women in the genetics condition ($M = 12.60$, $SD = 2.67$) were significantly lower than the math test scores of women in the socialization condition ($M = 16.00$, $SD = 3.59$), $t(18) = −2.40$, $p = .03$, $d = 1.07$, 95% CI [−6.37, −0.43]. These results suggest that statements that women have a genetic disadvantage adversely affect their math test performance.[2]

---

[2] While the data presented in this guide are all fabricated, researchers at the University of British Columbia conducted a similar study and found results consistent with these. The reference for the paper is: Dar-Nimrod, I., & Heine, S. J. (2006). Exposure to scientific theories affects women's math performance. *Science, 314*, 435. doi: 10.1126/science.1131100

Calculating the 99% Confidence Intervals for the Difference in Means

Once again, by default, SPSS provides the 95% confidence interval for the difference in means. However, this setting can easily be changed to produce any confidence interval you'd like. We'll practice by changing the setting to the 99% confidence interval.

Go to **Analyze→Compare Means→Independent-Samples T Test.** The "Independent-Samples T Test" dialogue window should still show the dependent variable—MathScore—in the Test Variables: box and the independent variable—Message—in the Grouping Variable: box. The groups should still be defined as 1 and 2. Simply click the **Options box** in the "Independent-Samples T Test" dialogue window to open the "Independent-Samples T Test: Options" dialogue window shown below. Replace the value of 95 with a value of **99** by simply typing in the box. Press **Continue** to close the dialogue window and press **OK** to run the analysis.

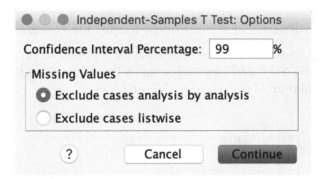

The same tables will appear in the output window. The only difference between these tables and the ones produced previously are that the "Independent Samples Effect Sizes" table now displays the 99% confidence interval for the three effect-size indicators and the "Independent Samples Test" table now displays the 99% confidence interval for the difference in means. According to the 99% confidence interval for the difference in means, we can be 99% certain that the interval -7.48 to 0.68 contains the true effect of the statement that women are genetically predisposed to be poor at math on women's actual math test performance. Since the confidence interval crosses 0 (and the effect-size estimator confidence intervals also cross 0), we must conclude that the effect is not statistically significant. Note that we would have reached the same conclusion using the traditional p value method and an alpha of .01 (the alpha level that corresponds to the 99% confidence interval) because $p = .03 > .01$.

Independent Samples Test

		Levene's Test for Equality of Variances		t-test for Equality of Means					99% Confidence Interval of the Difference	
		F	Sig.	t	df	Sig. (2–tailed)	Mean Difference	Std. Error Difference	Lower	Upper
MathScore	Equal variances assumed	.604	.447	-2.401	18	.027	-3.40000	1.41578	-7.47525	.67525
	Equal variances not assumed			-2.401	16.639	.028	-3.40000	1.41578	-7.51431	.71431

Reporting the Results

The results using an alpha of .01 and the 99% confidence interval could be reported in the following manner:

> Using an alpha of .01, an independent-samples t-test revealed that the math test scores of women in the genetics condition ($M = 12.60$, $SD = 2.67$) were not significantly different than the math test scores of women in the socialization condition ($M = 16.00$, $SD = 3.59$), $t(18) = -2.40$, $p = .03$, $d = 1.07$, 99% CI $[-7.48, 0.68]$.

ONE-TAILED INDEPENDENT-SAMPLES *t*-TEST

Now assume you had originally made a directional hypothesis and predicted that the message that women are genetically predisposed to be poor at math would *decrease* their math test performance. In this scenario, you would need to conduct a one-tailed independent-samples *t*-test. The method of conducting the analysis would remain the same, and the tables in the output would remain the same. The only thing that would change in this scenario is the reported *p* value, which would once again be cut in half. The *p* value for the two-tailed test was .0273, so the *p* value for a one-tailed test would be .01 (.0273 ÷ 2 = .0137). Since the mean of the genetics condition was lower than the mean of the socialization condition, we know the results are in the predicted direction.

Reporting the Results

We could report the results of a one-tailed independent-samples *t*-test (with alpha of .05) as follows:

A one-tailed independent-samples *t*-test revealed that the math test scores of women in the genetics condition ($M = 12.60$, $SD = 2.67$) were significantly lower than the math test scores of women in the socialization condition ($M = 16.00$, $SD = 3.59$), $t(18) = -2.40$, $p = .01$ (one-tailed), $d = 1.07$.

Predicting the Wrong Direction

Once again, if the results had worked out in the opposite direction we predicted and showed that math test scores were actually higher in the genetics condition, then we would need to make a further adjustment to the *p* value. Rather than simply dividing the *p* value in half as we did when the results worked out in the predicted direction, we would now need to subtract the divided *p* value from a value of 1. So, we would still divide the *p* value in half (.0273 ÷ 2 = .0137), but now we would also need to subtract the result from a value of 1 (1 − .0137 = .99) since in this scenario the results are opposite to those we predicted. In this case, we would report the following: $t(18) = 2.40$, $p = .99$ (one-tailed), $d = 1.07$. Since $p > .05$ in this scenario, we would need to conclude that statements that women have a genetic disadvantage do not adversely affect their math test performance.

One-Way Anova

Learning
Objectives

In this chapter, you will learn how to analyze data using one-way between-subjects ANOVA and one-way within-subjects ANOVA. For each, you will learn how to conduct the analysis and interpret and report the results. You will also learn how to calculate and interpret the effect size indicator eta-squared and perform follow up post hoc tests.

INTRODUCTION TO ONE-WAY ANOVA

As reviewed in Chapter Eight, *t*-tests are used to compare the means of *two* groups. One-way ANOVA is simply an extension of the *t*-test; it is used to compare the means of *three or more* groups. In this chapter, we will first consider one-way between-groups ANOVA, which is used in conjunction with between subjects (independent groups) designs, in which the various conditions contain different groups of participants. We will then consider one-way within-groups ANOVA, which is used to analyze data from within subjects (repeated measures) designs, in which the various conditions contain the same participants.

ONE-WAY BETWEEN-GROUPS ANOVA

Once again, we will begin with a one-way between-groups ANOVA. Assume you learned about the results of the study, described in Chapter Eight, on the influence of statements about women's math abilities on their math test performance. You decide that you could improve on the study by including a control condition in which women overhear a message unrelated to math abilities.

You invite 30 women to your lab and you randomly assign each woman to one of three separate conditions. Women in each of the conditions are told to wait in a waiting area where they "accidentally overhear" a staged conversation between two professors. Women in condition 1 (the genetics condition) overhear the professors discussing a new paper revealing evidence that a math gene has been discovered that is sex-linked and commonly deficient in women and that this may account for why women are underrepresented in the sciences. Women in condition 2 (the socialization condition) overhear the professors discussing a new paper revealing evidence that women tend to be underrepresented in the sciences only because they are raised in a society where they are regarded as being inferior at math. Women in condition 3 (the control condition) overhear the professors discussing their plans for the weekend. Participants are then asked to complete a math test containing 20 questions. You hypothesize that women's math test performance will be affected by the message they overhear. Prior to collecting any data, you decide to set alpha at .05. Note that ANOVA always uses a two-tailed test and it is not appropriate to use a one-tailed ANOVA. This is because the *F* statistic is a ratio of variances and, as such, will always be positive. Therefore, the *F* distribution is positively skewed rather than two-tailed.

Assume you record the following data. Once again, the codes "1," "2," and "3" under the columns labeled "Message" indicate which condition participants were assigned to, and the values in the "MathScore" columns indicate how many math questions each woman got correct.

ID	Message	MathScore	ID	Message	MathScore	ID	Message	MathScore
1	1	12	11	2	20	21	3	18
2	1	11	12	2	15	22	3	9
3	1	15	13	2	11	23	3	14
4	1	15	14	2	20	24	3	15
5	1	14	15	2	17	25	3	9
6	1	14	16	2	20	26	3	11
7	1	12	17	2	20	27	3	14
8	1	8	18	2	15	28	3	15
9	1	9	19	2	20	29	3	15
10	1	16	20	2	19	30	3	20

Begin by **entering** the **data** into a **blank SPSS spreadsheet**. Note that the top row contains the variable names, which must be entered in the Variable View window. While you are in Variable View, use the **Values** column to identify message **1** as **Genetics**, message **2** as **Socialization**, and message **3** as **Control**. Set the scales of measure to **Nominal** for **ID** and **Message** and to **Scale** for **MathScore**. The remaining data must be entered into three columns and 30 rows in the Data View window. Remember different variables must be entered into separate columns, whereas the data for each of the participants must be entered in the rows. Once the data are entered, your Data View window should look like the one displayed here.

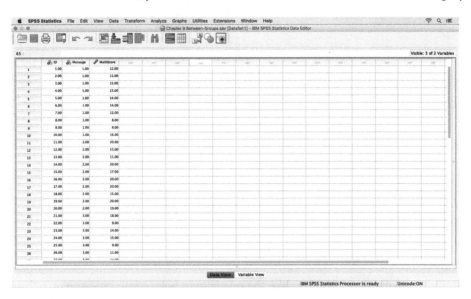

Conducting a One-Way Between-Groups ANOVA (Analyze→General Linear Model→Univariate)

To conduct a one-way between-groups ANOVA, you need to use the upper toolbar to go to **Analyze→General Linear Model→Univariate**.[1]

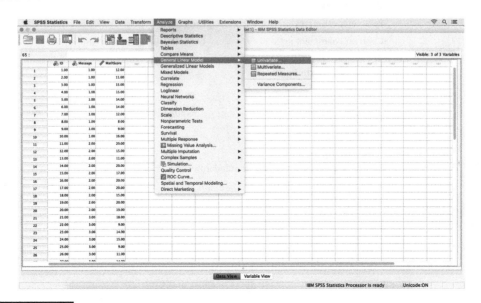

---

[1] The analysis can also be performed using the option Analyze→Compare Means→One-Way ANOVA; however, this method does not provide an option to compute an effect size indicator.

Next, a "Univariate" dialogue window will open. Highlight the dependent variable—**MathScore**—by clicking on the variable name, and then click the corresponding **blue arrow** to move it over to the **Dependent Variable: box**. Next, highlight the independent variable—**Message**—by clicking on the variable name, and move it over to the **Fixed Factor(s): box** by clicking the corresponding **blue arrow**. Next, click on the **Options...** tab.

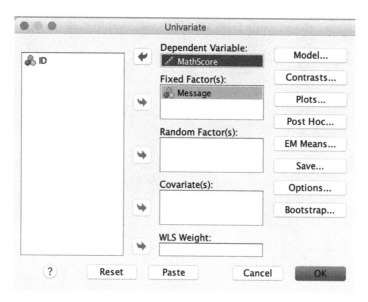

A "Univariate: Options" dialogue window, like the one shown below, will now appear. Check the options for **Descriptive statistics**, **Estimates of effect size**, and **Homogeneity tests**. Close the dialogue window by clicking **Continue**.

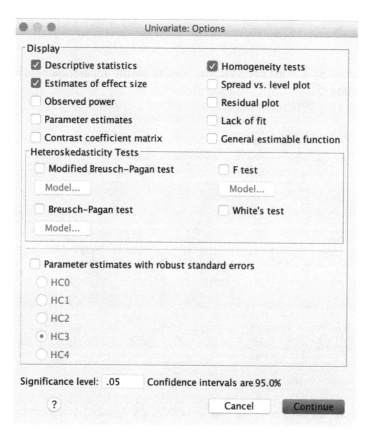

Next, click the **Post Hoc** option in the "Univariate" dialogue window. This will open the "Univariate: Post Hoc Multiple Comparisons for Observed Means" dialogue window, shown below. Move the independent variable **Message** over to the **Post Hoc Tests for: box** by clicking on the **blue arrow**. **Check** the **Bonferroni** and **Tukey** options, and then click **Continue** to close the dialogue window. Finally, click **OK** on the "Univariate" dialogue window to execute the analysis.

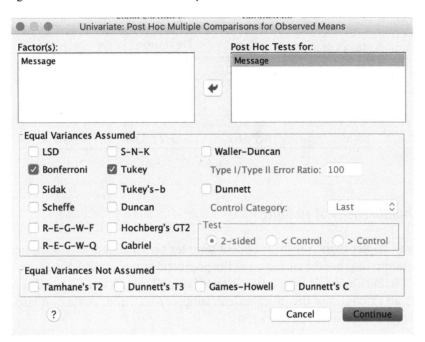

Interpreting the Results

An output window will now appear, displaying a series of tables. We will begin by examining the "Between-Subjects Factors" table shown below. This table simply shows the labels for the three conditions, the codes used to designate each condition, and the number of participants in each condition. Since these results are for a between-subjects design, it is possible that the sample size (n) will be different across the conditions.

Between–Subjects Factors

		Value Label	N
Message	1.00	Genetics	10
	2.00	Socialization	10
	3.00	Control	10

Descriptive Statistics

The next table in the output displays the labels we entered for the three conditions in the column labeled with the name of the independent variable ("Message"), as well as the means, standard deviations, and sample sizes for each of the three conditions in the columns labeled "Mean," "Std. Deviation," and "N," respectively. Additionally, the bottom row labeled "Total" displays the overall mean, standard deviation, and sample size for all three conditions combined.

Descriptive Statistics

Dependent Variable: MathScore

Message	Mean	Std. Deviation	N
Genetics	12.6000	2.67499	10
Socialization	17.7000	3.12872	10
Control	14.0000	3.55903	10
Total	14.7667	3.73874	30

Levene's Test of Homogeneity of Variance

If you'll recall from Chapter Eight, the independent samples t-test carries the assumption of homogeneity of variance (the assumption that the variances of the populations from which the samples were drawn are not significantly different), and this assumption is tested using Levene's test. The one-way between-groups ANOVA is simply an extension of the independent samples t-test, and, as such, it also carries this same assumption that is also tested using Levene's test.

The table labeled "Levene's Test of Equality of Error Variances" displays the results of this test of the assumption of homogeneity of variance. If the p value displayed at the intersection of the row labeled "Based on Mean" and the column labeled "Sig." is .05 or lower, then the variances are significantly different, and the assumption has been violated. This violation should then be reported along with the main results of the ANOVA. If the p value displayed in the table is greater than .05, then the variances are not significantly different, and the assumption has been met. As you can see below, for our example the p value associated with Levene's test is .86, which is greater than .05; therefore, the assumption of homogeneity of variance has been met.

Levene's Test of Equality of Error Variances[a,b]

		Levene Statistic	df1	df2	Sig.
MathScore	Based on Mean	.154	2	27	.858
	Based on Median	.087	2	27	.917
	Based on Median and with adjusted df	.087	2	22.260	.917
	Based on trimmed mean	.149	2	27	.862

Tests the null hypothesis that the error variance of the dependent variable is equal across groups.

a. Dependent variable: MathScore

b. Design: Intercept + Message

Main Effect

The primary results of the analysis are displayed in the table labeled "Tests of Between-Subjects Effects" shown below. The column labeled "Type III Sum of Squares" provides the values of the sums of squares (the sum of the squared deviation scores). The column labeled "Mean Square" provides the values of the mean squares (estimates of the population variances). The column labeled "df" contains the degrees of freedom, the column labeled "F" provides the *F* statistics, the column labeled "Sig." lists the *p* values, and the column labeled "Partial Eta Squared" provides the effect size estimates.

Tests of Between–Subjects Effects

Dependent Variable: MathScore

Source	Type III Sum of Squares	df	Mean Square	F	Sig.	Partial Eta Squared
Corrected Model	138.867[a]	2	69.433	7.035	.003	.343
Intercept	6541.633	1	6541.633	662.755	.000	.961
Message	138.867	2	69.433	7.035	.003	.343
Error	266.500	27	9.870			
Total	6947.000	30				
Corrected Total	405.367	29				

a. R Squared = .343 (Adjusted R Squared = .294)

The *F* statistic is the primary statistic considered in ANOVA. It is a ratio of two types of variance. Specifically, it is the ratio of between groups variance over within groups variance. Between groups variance reflects the variability of the means *across* the groups, while within groups variance reflects the variability of scores *within* each group. Within groups variance is error variance (i.e., variability in the dependent variable that is unrelated to the independent variable), and, as such, SPSS provides the statistics related to it in the row labeled "Error." Between groups variance essentially reflects the effect of the independent variable on the dependent variable, and, as such, the statistics related to it are provided in the row labeled with the name of the independent variable. We named our independent variable message; therefore, the row labeled "Message" in the table shown above indicates that the between groups sum of squares is 138.87, the between groups degrees of freedom are 2, and the mean square between is 69.43. This mean square between value is computed by dividing the sum of squares by the degrees of freedom (138.8667/2 = 69.43). The row labeled "Error" shows that the within groups sum of squares is 266.50, the within groups degrees of freedom are 27, and the mean square within is 9.87. The mean square within value is also computed by dividing the sum of squares by the degrees of freedom (266.5000/27 = 9.87). The value of *F* is then computed by dividing the mean square between value by the mean square within (69.4334/9.8704 = 7.03).

The main effect is the overall effect of the independent variable on the dependent variable. Statistics related to this main effect are provided in the row labeled with the name of the independent variable. Therefore, the primary statistics of interest to us are provided in the row labeled "Message." As displayed in that row, the value of *F* is 7.03, the *p* value is .003, and the effect size is .34. As alluded to earlier, there are two degrees of freedom of interest that are associated with the *F* test: between groups degrees of freedom and within groups degrees of freedom. The value of the between groups degrees of freedom is listed in the row labeled with the name of the independent variable, while the value of the within groups degree of freedom is listed in the row labeled "Error." For all ANOVAs, these degrees of freedom are reported in parentheses, separated by a comma. Putting it all together, the main effect of the message women overheard on their math test scores would be reported as: $F(2, 27) = 7.03, p = .003, \eta_p^2 = .34$. These results suggest that there is a significant effect of the message women overheard on their math test performance.

Effect Size

One of the most commonly used effect size indicators for ANOVA is partial eta-squared, symbolized as η_p^2. Because it is a Greek character, it should not be italicized. Moreover, since partial eta-squared is a proportion whose values cannot exceed 1, leading 0s before the decimal are not appropriate to use. Partial eta-squared values of .01, .06, and .14 are considered small, medium, and large, respectively. Partial eta-squared is also sometimes referred to as R^2 because, like R^2, it indicates the proportion of variability in the dependent variable that is accounted for by the independent variable. Indeed, in addition to displaying the eta-squared value for the main effect of the independent variable in the column labeled "Partial Eta Squared," the value is also provided under the table with the label "R Squared." As shown both in and under the table, the eta-squared value of interest to us is .34. This is a large effect and indicates that the message women overheard accounts for 34.26% of the variability in their math test scores.

Post Hoc Tests

A significant main effect (F statistic) indicates that at least one of the means differs from at least one of the other means. However, the F statistic does not provide any indication of which means differ significantly from which. In order to determine which of the means differ significantly from each other we need to use post hoc tests. Thus, post hoc tests are necessary (and in most cases appropriate) only when the overall main effect is statistically significant. This is because if the F statistic is not significant, then it indicates that none of the means differ significantly from one another and any follow-up analysis using post hoc tests could be seen as a fishing exhibition. Since our main effect was statistically significant, we will need to consider the results of the post hoc tests to determine which means differ significantly. The results of the post hoc comparisons are displayed in the following table.

Multiple Comparisons

Dependent Variable: MathScore

	(I) Message	(J) Message	Mean Difference (I–J)	Std. Error	Sig.	95% Confidence Interval Lower Bound	95% Confidence Interval Upper Bound
Tukey HSD	Genetics	Socialization	−5.1000*	1.40502	.003	−8.5836	−1.6164
		Control	−1.4000	1.40502	.585	−4.8836	2.0836
	Socialization	Genetics	5.1000*	1.40502	.003	1.6164	8.5836
		Control	3.7000*	1.40502	.036	.2164	7.1836
	Control	Genetics	1.4000	1.40502	.585	−2.0836	4.8836
		Socialization	−3.7000*	1.40502	.036	−7.1836	−.2164
Bonferroni	Genetics	Socialization	−5.1000*	1.40502	.004	−8.6862	−1.5138
		Control	−1.4000	1.40502	.984	−4.9862	2.1862
	Socialization	Genetics	5.1000*	1.40502	.004	1.5138	8.6862
		Control	3.7000*	1.40502	.041	.1138	7.2862
	Control	Genetics	1.4000	1.40502	.984	−2.1862	4.9862
		Socialization	−3.7000*	1.40502	.041	−7.2862	−.1138

Based on observed means.
 The error term is Mean Square(Error) = 9.870.

*. The mean difference is significant at the

The top section of the table, labeled "Tukey HSD" contains the results of Tukey's Honestly Significant Difference (*HSD*) post hoc test. The bottom section of the table, labeled "Bonferroni" contains the results of the Bonferroni post hoc test. Both are respected and commonly used post hoc tests that control for the inflation in Type I error that could occur with multiple comparisons (i.e., they keep the overall alpha level for the series of tests at .05). The Bonferroni test tends to be a bit more conservative than Tukey's *HSD* test (so Type I errors are less likely, but Type II errors become more likely with the Bonferroni test). Typically, only one post hoc analysis is performed and reported. Two types are reported here only to offer a description and demonstration of each of these commonly used tests.

The columns labeled "(I) Message" and "(J) Message" display the conditions being compared. The column labeled "Mean Difference (I − J)" shows the differences in the means of the two conditions being compared. The column labeled "Std. Error" contains the standard error of the means. The column labeled "Sig." contains the *p* values for the comparisons, and the final columns contain the 95% confidence intervals for the difference in the means.

Let's start with the results of Tukey's *HSD* test shown in the upper portion of the table. The first row contains a comparison of the genetics and socialization conditions. The value of −5.10 in the column labeled "Mean Difference (I − J)" indicates that the mean of the genetics condition is 5.10 units lower than the mean of the socialization condition. The *p* value of .003 displayed in the "Sig." column indicates that this difference is statistically significant. The next row contains the comparison of the genetics and control conditions. The *p* value of .59 listed in this row indicates that the difference of −1.40 in the means of these two conditions is not statistically significant. The next row comparing the socialization and genetics conditions is redundant with the first row. However, the following row, comparing the socialization and control conditions reveals that the difference of 3.70 in the means of these two conditions is statistically significant (*p* = 04). The final two rows of the upper portion of the table are redundant with the second and fourth rows that we reviewed.

A review of the bottom portion of the "Multiple Comparisons" table reveals that the results of the comparisons using the Bonferroni procedure are consistent with those of Tukey's *HSD* test. Specifically, they show that the mean difference of −5.10 across the genetics and socialization conditions is statistically significant (*p* = .004), that the difference of −1.40 in the means of the genetics and control conditions is not statistically significant (*p* = .98), and that the mean difference of 3.70 across the socialization and control conditions is statistically significant (*p* = .04). Notice that the *p* values associated with the more conservative Bonferroni method are slightly higher than those associated with Tukey's *HSD* test.

Reporting the Results

Once again, we would typically only compute and report the results of one post hoc test. Therefore, the result of the one-way between-groups ANOVA with the Tukey's *HSD* post hoc test are reported below:

> A one-way between-groups ANOVA revealed a large-sized significant main effect of the message women overheard on their math test performance, $F(2, 27) = 7.03$, $p = .003$, $\eta_p^2 = .34$. Post hoc comparisons using Tukey's *HSD* test indicated that the math test scores of women in the genetics ($M = 12.60$, $SD = 2.67$) condition were significantly lower than the math test scores of women in the socialization ($M = 17.70$, $SD = 3.13$) condition ($p = .003$). Women in the control condition also had significantly lower math test scores ($M = 14.00$, $SD = 3.56$) than those in the socialization condition ($p = .04$). The difference between the math test scores of women in the genetics and control conditions was not statistically significant ($p = .59$).

ONE-WAY WITHIN-GROUPS ANOVA

Whereas one-way between-groups ANOVA is used in conjunction with between subjects designs with three or more groups (i.e., to compare the means of three or more groups of different participants), one-way within-groups ANOVA is used in conjunction with within subjects (repeated measures) designs with three or more conditions. Specifically, it is used to determine whether the means of three or more conditions (all containing the same participants) differ significantly.

Assume you learned about the results of the study described in Chapter Eight on the acute effects of THC on memory test performance, and you want to further examine whether the acute effects of THC differ from those of alcohol, in addition to the placebo control group. You hypothesize that performance in the THC and alcohol conditions will be different than performance in the placebo condition. Before conducting the study, you decide to set alpha at .05. Assume that you randomly select and invite 24 individuals to your lab on three separate days. Each individual is given a drink containing cannabis on one day, a drink containing alcohol on another day, and a placebo drink that doesn't contain any THC or alcohol on another day. To guard against order effects, you use complete counterbalancing. After consuming each substance, participants are given a memory test. They are read a list of 15 words and are asked to recall as many words from the list as possible. You measure the number of words each participant correctly recalls. Assume you obtain the data shown below. They depict the mean number of words that each participant correctly recalled after consuming each substance.

ID	THC	Alcohol	Placebo	ID	THC	Alcohol	Placebo
1	7	6	8	13	6	6	8
2	5	5	7	14	5	7	7
3	4	3	5	15	9	8	12
4	12	13	10	16	10	11	12
5	8	6	9	17	15	15	15
6	9	4	6	18	7	9	11
7	11	13	14	19	5	6	7
8	6	4	5	20	9	7	10
9	7	7	9	21	10	9	9
10	6	5	8	22	5	6	8
11	8	9	11	23	9	9	9
12	7	4	8	24	11	13	15

Begin by **entering** the **data** into a **blank SPSS spreadsheet**. Note that the top row contains the variable names, which must be entered in the Variable View window. The remaining data must be entered into four columns and 24 rows in the Data View window. Keep in mind that different variables must be entered into separate columns, whereas the data for each of the participants must be entered into the rows. (Therefore, each row must represent one participant's data on all of the variables.) Define the scales of measure to **Nominal** for **ID** and to **Scale** for **THC, Alcohol, and Placebo**. Once the data are entered, your Data View window should look like the one displayed here.

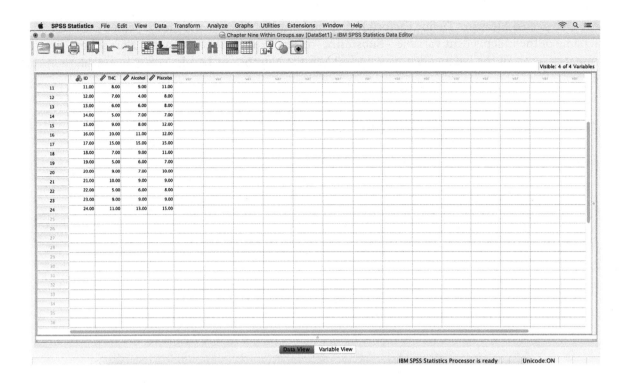

Conducting a One-Way Within-Groups ANOVA (Analyze→General Linear Model→Repeated Measures)

To conduct a one-way within-groups ANOVA, use the upper toolbar to go to **Analyze→General Linear Model→Repeated Measures.**

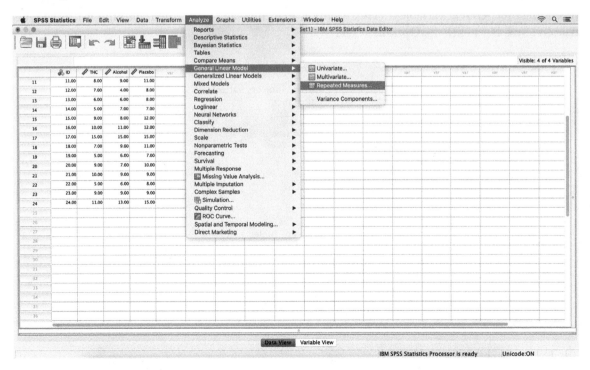

Next a "Repeated Measures Define Factor(s)" dialogue window, like the one shown below, will open. Type the name of the independent variable—**Substance**—in the **Within-Subject Factor Name: box**. Type the number of levels of the independent variable—**3**—in the **Number of Levels: box**. Click **Add,** then click **Define**.

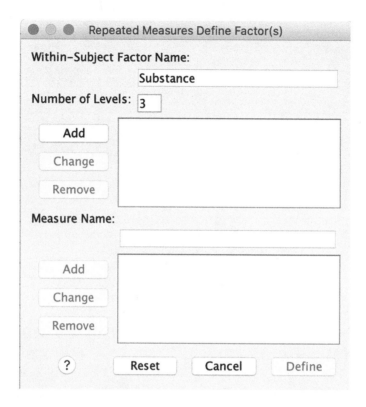

A "Repeated Measures" dialogue window like the one shown below will now appear. Highlight the three levels of the independent variable—**THC, Alcohol, Placebo**—and move them into the **Within-Subjects Variables: box**. Next click the **Options...** tab.

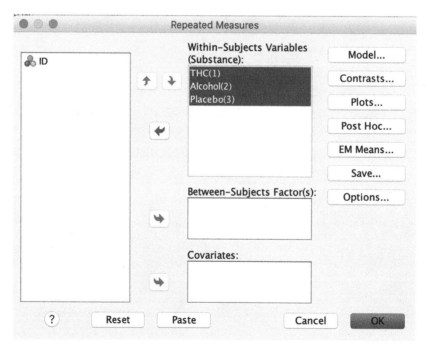

A "Repeated Measures: Options" dialogue window like the one shown below will now appear. Check the options for **Descriptive statistics** and **Estimates of effect size**. Close the dialogue window by clicking **Continue**. Note that if you want to use a more conservative alpha level (e.g., .01) and compute more stringent confidence intervals (e.g., 99%), you can change the alpha level by replacing the .05, in the Significance level: box at the bottom of the dialogue window, to the desired alpha level. We will, however, stick with the conventional .05 alpha level for this example.

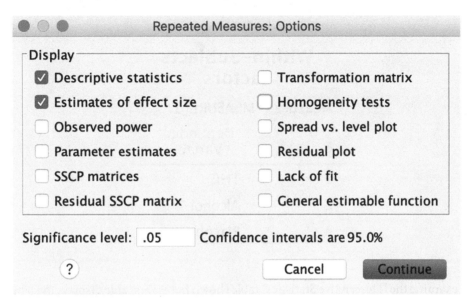

Next, click the **EM Means** option in the "Repeated Measures" dialogue window. This will open the "Repeated Measures: Estimated Marginal Means" dialogue window shown below. Move the independent variable **Substance** over to the **Display Means for: box** by highlighting the variable name in the "Factor(s) and Factor Interactions:" box on the left side and clicking on the **blue arrow**. **Check** the **Compare main effects** option, and then, using the drop-down menu, change the default from *LSD* (which is a liberal post hoc test that does not adjust for inflation to Type I error) to **Sidak,** which is a moderately conservative post hoc test. Specifically, Sidak is less conservative than Bonferroni, but more conservative than *LSD*, in the manner in which it adjusts for the inflation to Type I error rate that could result from performing multiple follow-up comparisons of the three groups. Finally click **Continue,** and then click **OK** on the main "Repeated Measures" dialogue window to execute the analysis.

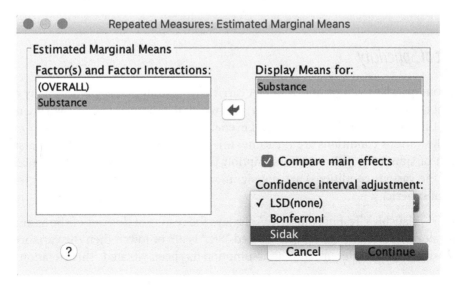

Interpreting the Results

Descriptive Statistics

An output window will now appear, displaying a series of tables. The first table simply displays the names of the three conditions and the numbers used to code each condition. These numbers will be useful when we interpret the post hoc comparisons later in this chapter.

Within-Subjects Factors

Measure: MEASURE_1

Substance	Dependent Variable
1	THC
2	Alcohol
3	Placebo

Next, we will examine the "Descriptive Statistics" table shown here. This table displays the labels we entered for the three groups in the first column, as well as the means, standard deviations, and sample sizes for each of the three conditions in the columns labeled "Mean," "Std. Deviation," and "N," respectively.

Descriptive Statistics

	Mean	Std. Deviation	N
THC	7.9583	2.66179	24
Alcohol	7.7083	3.29003	24
Placebo	9.2917	2.80495	24

Mauchly's Test of Sphericity

If you'll recall, one-way between-groups ANOVA carries the assumption of homogeneity of variance (the assumption that the variances of the populations from which the samples were drawn are not significantly different), and this assumption is tested using Levene's test. Since for one-way within-groups ANOVA the people in the various conditions are the same, this test carries a slightly different assumption called the assumption of sphericity. This is the assumption that the variances of the difference scores (differences between the various conditions) are not significantly different. This assumption is assessed using Mauchly's test of sphericity.

The table labeled "Mauchly's Test of Sphericity," displays the results of this test of the assumption of sphericity. If the p value displayed in the column labeled "Sig." is .05 or lower, then the variances of the difference scores are significantly different, and the assumption has been violated. This violation should then be reported along with the Greenhouse–Geisser adjusted results of the ANOVA. If the p value displayed in the table is greater than .05, then the variances are not significantly different, and the assumption has been met. For our example, the p value associated with Mauchly's test of sphericity test is .62, which is greater than .05; therefore, the assumption has been met.

Mauchly's Test of Sphericity[a]

Measure: MEASURE_1

Within Subjects Effect	Mauchly's W	Approx. Chi-Square	df	Sig.	Epsilon[b] Greenhouse–Geisser	Huynh–Feldt	Lower-bound
Substance	.957	.962	2	.618	.959	1.000	.500

Tests the null hypothesis that the error covariance matrix of the orthonormalized transformed dependent variables is proportional to an identity matrix.

a. Design: Intercept
   Within Subjects Design: Substance

b. May be used to adjust the degrees of freedom for the averaged tests of significance. Corrected tests are displayed in the Tests of Within–Subjects Effects table.

Main Effect

The primary results of the analysis are displayed in the table labeled "Tests of Within-Subjects Effects". The column labeled "Type III Sum of Squares" provides the values of the sums of squares (the sum of the squared deviation scores). The column labeled "Mean Square" provides the values of the mean squares (estimates of the population variances). The column labeled "df" contains the degrees of freedom, the column labeled "F" provides the F statistics, the column labeled "Sig." lists the p values, and the column labeled "Partial Eta Squared" provides the effect size estimates.

Once again, the F statistic is the primary statistic considered in ANOVA. It is a ratio of two types of variance. It is a ratio of between groups variance to within groups variance. Once again, between groups variance reflects the variability of the means *across* the conditions, and within groups variance reflects the variability of scores *within* each condition. Within groups variance is essentially error variance (i.e., it is variance that is unrelated to the manipulation of the independent variable), and, as such, SPSS provides the statistics related to it in the row labeled "Error." Between groups variance essentially reflects the effect of the independent variable on the dependent variable, and, as such, the statistics related to it are provided in the row labeled with the name of the independent variable. Since we did not violate the assumption of sphericity, we can refer to the rows labeled "Sphericity Assumed[2]." We named our independent variable Substance; therefore, the row labeled "Substance" in the table indicates that the between groups sum of squares is 34.78, the between groups degrees of freedom are 2, and the mean square between is 17.39. Recall that the mean square between value is computed by dividing the sum of squares by the degrees of freedom (34.7778/2 = 17.39). The row labeled "Error" shows that the within groups sum of squares is 65.22, the within groups degrees of freedom are 46, and the mean square within is 1.42. Again, the mean square within value is computed by dividing the sum of squares by the degrees of freedom (65.2222/46 = 1.42). The F statistic is then computed by dividing the mean square between by the mean square within (17.3890/1.4179 = 12.26).

---

[2] If we had of violated the assumption of sphericity then we would need to report the values in the rows labeled "Greenhouse-Geisser" as these contain statistics that have been adjusted to help protect against the inflation in Type I error that can occur when the assumption is violated.

Tests of Within–Subjects Effects

Measure: MEASURE_1

Source		Type III Sum of Squares	df	Mean Square	F	Sig.	Partial Eta Squared
Substance	Sphericity Assumed	34.778	2	17.389	12.264	.000	.348
	Greenhouse–Geisser	34.778	1.918	18.133	12.264	.000	.348
	Huynh–Feldt	34.778	2.000	17.389	12.264	.000	.348
	Lower–bound	34.778	1.000	34.778	12.264	.002	.348
Error(Substance)	Sphericity Assumed	65.222	46	1.418			
	Greenhouse–Geisser	65.222	44.113	1.479			
	Huynh–Feldt	65.222	46.000	1.418			
	Lower–bound	65.222	23.000	2.836			

The main effect is the overall effect of the independent variable on the dependent variable. Once again, statistics related to the main effect of the independent variable on the dependent variable are provided in the row labeled with the name of the independent variable. As displayed in the row labeled "Substance," the value of F is 12.26, the p value is less than .001, and the effect size is .35. As described previously, there are two degrees of freedom of interest that are associated with the F test, between groups degrees of freedom and within groups degrees of freedom. The value of the between groups degrees of freedom is listed in the row labeled with the name of the independent variable, whereas the value of the within groups degrees of freedom is listed in the row labeled "Error." The degrees of freedom are reported in parentheses, separated by a comma. Putting it all together, the main effect of the substance ingested on memory test scores would be reported as $F(2, 46) = 12.26$, $p < .001$, $\eta_p^2 = .35$. These results suggest that the manipulation of substances had a significant effect on memory test performance.

Effect Size

Once again, one of the most commonly used effect size indicators for ANOVA is partial eta-squared, symbolized as η_p^2. Recall that it is an indicator of the proportion of variability in the dependent variable accounted for by the independent variable. Eta is a Greek character, so it should not be italicized. Moreover, since the value cannot exceed 1, leading 0s before the decimal are not appropriate to use.

The partial eta-squared value reported in the "Tests of Within-Subjects Effects" table is .35. Recall that partial eta-squared values of .01, .06, and .14 are considered small, medium, and large, respectively. Therefore, there is a large-sized effect that indicates that the substance ingested accounts for 34.78% of the variability in memory test performance.

Post Hoc Tests

A significant main effect (F statistic) indicates that at least one of the means differs from at least one of the other means. However, the F statistic does not provide any indication of which means differ significantly from which. In order to determine which of the means differ significantly from each other, we need to use post hoc tests. Thus, post hoc tests are necessary (and in most cases appropriate) only when the overall main effect is statistically significant. This is because if the F statistic is not significant, then it indicates that none of the means differ significantly from one another. Since our main effect was statistically significant, we will need to consider the results of the post hoc tests to determine which means differ significantly. The results of the post hoc comparisons are displayed in the "Pairwise Comparisons" table shown here.

Pairwise Comparisons

Measure: MEASURE_1

(I) Substance	(J) Substance	Mean Difference (I–J)	Std. Error	Sig.[b]	95% Confidence Interval for Difference[b]	
					Lower Bound	Upper Bound
1	2	.250	.357	.868	–.670	1.170
	3	–1.333[*]	.364	.004	–2.272	–.395
2	1	–.250	.357	.868	–1.170	.670
	3	–1.583[*]	.306	.000	–2.372	–.794
3	1	1.333[*]	.364	.004	.395	2.272
	2	1.583[*]	.306	.000	.794	2.372

Based on estimated marginal means

*. The mean difference is significant at the

b. Adjustment for multiple comparisons: Sidak.

The columns labeled "(I) Substance" and "(J) Substance" display the conditions being compared. The column labeled "Mean Difference (I − J)" shows the differences in the means of the two conditions being compared. Unfortunately, SPSS displays the condition codes rather than names, but once again a legend displaying the condition codes and names is provided in the first table in the output. It shows that 1 = THC, 2 = alcohol, and 3 = placebo. The column labeled "Std. Error" contains the standard error of the mean differences. The column labeled "Sig." contains the p values for the comparisons, and the final columns contain the 95% confidence intervals for the mean differences.

The first row contains a comparison of the THC (1) and alcohol (2) conditions. The value of 0.25 in the column labeled "Mean Difference (I − J)," indicates that the mean of the THC condition is 0.25 units higher than the mean of the alcohol condition. The p value of .87 displayed in the "Sig." column indicates that this difference is not statistically significant. The next row contains the comparison of the THC (1) and placebo (3) conditions. The value of −1.33 indicates that participants recalled 1.33 fewer words, on average, in the THC condition relative to the placebo condition. The p value of .004 listed in this row indicates that this mean difference is statistically significant. The next row comparing the alcohol (2) and THC (1) conditions is redundant with the first row. However, the following row, comparing the alcohol (2) and placebo (3) conditions, reveals participants recalled on average 1.58 words fewer in the alcohol condition than the placebo condition. The p value is less than .001 which indicates that this difference is statistically significant. The final two rows of the table are redundant with the second and fourth rows that we reviewed.

Reporting the Results

The results of the one-way within-groups ANOVA are reported below:

A one-way within-groups ANOVA revealed a large-sized significant main effect of the substance ingested on memory test performance, $F(2, 46) = 12.26$, $p < .001$, $\eta_p^2 = .35$. Post hoc comparisons using Sidak's test indicated that the memory test scores in the THC condition ($M = 7.96$, $SD = 2.66$) were significantly lower than the memory test scores in the placebo control condition ($M = 9.29$, $SD = 2.80$) ($p = .004$). Similarly, memory test scores in the alcohol condition ($M = 7.71$, $SD = 3.29$) were significantly lower than those in the placebo control condition ($p < .001$). There was not a significant difference in the memory test scores in the THC and alcohol conditions ($p = .87$).